ANALYSIS OF ENGINEERING DESIGN STUDIES FOR DEMILITARIZATION OF ASSEMBLED CHEMICAL WEAPONS AT PUEBLO CHEMICAL DEPOT

Committee on Review and Evaluation of Alternative Technologies for
Demilitarization of Assembled Chemical Weapons: Phase II

Board on Army Science and Technology
Division on Engineering and Physical Sciences
National Research Council

NATIONAL ACADEMY PRESS
Washington, D.C.

NATIONAL ACADEMY PRESS   2101 Constitution Avenue, N.W.   Washington, DC 20418

NOTICE: The project that is the subject of this report was approved by the Governing Board of the National Research Council, whose members are drawn from the councils of the National Academy of Sciences, the National Academy of Engineering, and the Institute of Medicine. The members of the committee responsible for the report were chosen for their special competences and with regard for appropriate balance.

This is a report of work supported by Contract DAAD19-00-C-0009 between the U.S. Army and the National Academy of Sciences. Any opinions, findings, conclusions, or recommendations expressed in this publication are those of the author(s) and do not necessarily reflect the view of the organizations or agencies that provided support for the project.

International Standard Book Number 0-309-07607-2

Limited copies are available from:

Board on Army Science and Technology
National Research Council
2101 Constitution Avenue, N.W.
Washington, DC 20418
(202) 334-3118

Additional copies are available from:

National Academy Press
2101 Constitution Avenue, N.W.
Lockbox 285
Washington, DC 20055
(800) 624-6242 or (202) 334-3313
(in the Washington metropolitan area)
http://www.nap.edu

Copyright 2001 by the National Academy of Sciences. All rights reserved.

Printed in the United States of America

THE NATIONAL ACADEMIES

National Academy of Sciences
National Academy of Engineering
Institute of Medicine
National Research Council

The **National Academy of Sciences** is a private, nonprofit, self-perpetuating society of distinguished scholars engaged in scientific and engineering research, dedicated to the furtherance of science and technology and to their use for the general welfare. Upon the authority of the charter granted to it by the Congress in 1863, the Academy has a mandate that requires it to advise the federal government on scientific and technical matters. Dr. Bruce M. Alberts is president of the National Academy of Sciences.

The **National Academy of Engineering** was established in 1964, under the charter of the National Academy of Sciences, as a parallel organization of outstanding engineers. It is autonomous in its administration and in the selection of its members, sharing with the National Academy of Sciences the responsibility for advising the federal government. The National Academy of Engineering also sponsors engineering programs aimed at meeting national needs, encourages education and research, and recognizes the superior achievements of engineers. Dr. Wm. A. Wulf is president of the National Academy of Engineering.

The **Institute of Medicine** was established in 1970 by the National Academy of Sciences to secure the services of eminent members of appropriate professions in the examination of policy matters pertaining to the health of the public. The Institute acts under the responsibility given to the National Academy of Sciences by its congressional charter to be an adviser to the federal government and, upon its own initiative, to identify issues of medical care, research, and education. Dr. Kenneth I. Shine is president of the Institute of Medicine.

The **National Research Council** was organized by the National Academy of Sciences in 1916 to associate the broad community of science and technology with the Academy's purposes of furthering knowledge and advising the federal government. Functioning in accordance with general policies determined by the Academy, the Council has become the principal operating agency of both the National Academy of Sciences and the National Academy of Engineering in providing services to the government, the public, and the scientific and engineering communities. The Council is administered jointly by both Academies and the Institute of Medicine. Dr. Bruce M. Alberts and Dr. Wm. A. Wulf are chairman and vice chairman, respectively, of the National Research Council.

COMMITTEE ON REVIEW AND EVALUATION OF ALTERNATIVE TECHNOLOGIES FOR DEMILITARIZATION OF ASSEMBLED CHEMICAL WEAPONS: PHASE II

ROBERT A. BEAUDET, *Chair*, University of Southern California, Los Angeles
RICHARD J. AYEN, Waste Management, Inc. (retired), Wakefield, Rhode Island
JOAN B. BERKOWITZ, Farces Berkowitz and Company, Washington, D.C.
RUTH M. DOHERTY, Naval Surface Warfare Center, Indian Head, Maryland
WILLARD C. GEKLER, EQE International/PLG, Irvine, California
SHELDON E. ISAKOFF, E.I. du Pont de Nemours and Company (retired), Chadds Ford, Pennsylvania
HANK C. JENKINS-SMITH, University of New Mexico, Albuquerque
DAVID S. KOSSON, Vanderbilt University, Nashville, Tennessee
FREDERICK J. KRAMBECK, Exxon Mobil Research and Engineering Company, Fairfax, Virginia
JOHN A. MERSON, Sandia National Laboratories, Albuquerque, New Mexico
WILLIAM R. RHYNE, H&R Technical Associates, Inc., Oak Ridge, Tennessee
STANLEY I. SANDLER, University of Delaware, Newark
WILLIAM R. SEEKER, General Electric Energy and Environmental Research Corporation, Irvine, California
LEO WEITZMAN, LVW Associates, Inc., West Lafayette, Indiana

Board on Army Science and Technology Liaison

JOSEPH J. VERVIER, ENSCO, Inc., Indiatlantic, Florida

Staff

PATRICIA P. PAULETTE, Study Director
HARRISON T. PANNELLA, Program Officer
JACQUELINE CAMPBELL-JOHNSON, Senior Project Assistant
GWEN ROBY, Senior Project Assistant
JAMES C. MYSKA, Research Associate

BOARD ON ARMY SCIENCE AND TECHNOLOGY

WILLIAM H. FORSTER, *Chair*, Northrop Grumman Corporation, Baltimore, Maryland
JOHN E. MILLER, *Vice Chair*, Oracle Corporation, Reston, Virginia
ROBERT L. CATTOI, Rockwell International (retired), Dallas, Texas
RICHARD A. CONWAY, Union Carbide Corporation (retired), Charleston, West Virginia
GILBERT F. DECKER, Walt Disney Imagineering, Glendale, California
PATRICK F. FLYNN, Cummins Engine Company, Inc. (retired), Columbus, Indiana
HENRY J. HATCH, Chief of Engineers, U.S. Army (retired), Oakton, Virginia
EDWARD J. HAUG, University of Iowa, Iowa City
GERALD J. IAFRATE, North Carolina State University, Raleigh
MIRIAM E. JOHN, California Laboratory, Sandia National Laboratories, Livermore, California
DONALD R. KEITH, Cypress International (retired), Alexandria, Virginia
CLARENCE W. KITCHENS, IIT Research Institute, Alexandria, Virginia
KATHRYN V. LOGAN, Georgia Institute of Technology (professor emerita), Roswell, Georgia
JOHN W. LYONS, U.S. Army Research Laboratory (retired), Ellicott City, Maryland
JOHN H. MOXLEY III, Korn/Ferry International, Los Angeles, California
STEWART D. PERSONICK, Drexel University, Philadelphia, Pennsylvania
MILLARD F. ROSE, Radiance Technologies, Huntsville, Alabama
GEORGE T. SINGLEY III, Hicks and Associates, Inc., McLean, Virginia
CLARENCE G. THORNTON, Army Research Laboratories (retired), Colts Neck, New Jersey
JOHN D. VENABLES, Venables and Associates, Towson, Maryland
JOSEPH J. VERVIER, ENSCO, Inc., Indiatlantic, Florida

Staff

BRUCE A. BRAUN, Director
MICHAEL A. CLARKE, Associate Director
WILLIAM E. CAMPBELL, Administrative Coordinator
CHRIS JONES, Financial Associate
GWEN ROBY, Administrative Assistant
DEANNA P. SPARGER, Senior Project Assistant

Preface

The United States has been in the process of destroying its chemical munitions for well over a decade. Initially, the U.S. Army, with recommendations from the National Research Council (NRC), decided to use incineration as its destruction method at all sites. However, citizens in some states with stockpile storage sites have opposed incineration on the grounds that it is impossible to determine the exact nature of the effluents escaping from the stacks. Although the Army has continued to pursue incineration at four of the eight storage sites in the continental United States, in response to growing public opposition to incineration in Maryland and Indiana and a 1996 report by the NRC, *Review and Evaluation of Alternative Chemical Disposal Technologies*, the Army is developing alternative processes to neutralize chemical agents using hydrolysis. These processes will be used to destroy the VX nerve agent at Newport, Indiana, and the mustard agent at Aberdeen, Maryland, both of which are stored in bulk one-ton containers.

In 1996, persuaded by the public opposition in Lexington, Kentucky, and Pueblo, Colorado, Congress enacted Public Law 104-201, which instructed the Department of Defense (DOD) to "conduct an assessment of the chemical demilitarization program for destruction of assembled chemical munitions and of the alternative demilitarization technologies and processes (other than incineration) that could be used for the destruction of the lethal chemical agents that are associated with these munitions." The Army established a Program Manager for Assembled Chemical Weapons Assessment (PMACWA) to respond to Congress. In Public Law 104-208, the PMACWA was required to "identify and demonstrate not less than two alternatives to the baseline incineration process for the demilitarization of assembled chemical munitions." Following the demonstration of six technologies, the PMACWA selected two as candidates for destroying the weapons at Pueblo Chemical Depot. The two packages have since progressed to the engineering design phase of the Assembled Chemical Weapons Assessment (ACWA) program.

In contrast with prior chemical weapons demilitarization programs, the PMACWA has involved citizen stakeholders in every aspect of the program, including the procurement process. A nonprofit organization, the Keystone Center, was hired to facilitate public involvement through a process known as the Dialogue. The Dialogue group, whose 35 members represent the Army and various community stakeholders, developed the criteria for selecting the technologies and were involved in all other aspects of the selection process. The Dialogue process has become a model for public involvement in matters of public concern. Indeed, the Department of Energy and the National Aeronautics and Space Administration have also adopted this approach.

Congress mandated that the Army coordinate with the NRC during the ACWA program. In response, the NRC established the Committee on Review and Evaluation of Alternative Technologies for Demilitarization of Assembled Chemical Weapons (ACW I Committee) in 1997 to oversee this program. The issue before the committee was not whether incineration is an adequate technology but whether, given that some citizens are strongly opposed to that method, other chemical methods, acceptable to the stakeholders, could be used. The Committee on Review and Evaluation of Alternative Technologies for Demilitarization of Assembled Chemical Weapons: Phase II (ACW II Committee) was established in the spring of 2000 for the engineering design phase of the ACWA program.

One goal of this study is to provide an independent technical evaluation of the engineering-design packages of the two candidate processes being considered for use at the Pueblo Chemical Depot. This evaluation is expected to contribute to DOD's Record of Decision (ROD) for the selection of a technology for the Pueblo site. The ROD was scheduled to be released on August 30, 2001. Therefore, to be of value in the selections, this report had to be published by mid-July 2001. Unfortunately, not all of the tests associated with the two packages, which address all aspects of demilitarization from disassembly of the weapons to the disposal of waste streams, were completed at the time that data gathering for this report had to be terminated to meet the mid-July deadline.

I wish to express my gratitude to the members of the ACW II Committee, all of whom served as volunteers and many of whom served with me on the ACW I Committee. They have all given unselfishly of their time and knowledge. Committee members' areas of expertise include chemical processing, biological remediation, environmental regulations and permitting, energetic materials, and public acceptance. Each member attended plenary meetings, visited the headquarters of technology providers and test sites, observed design-review sessions, and studied the extensive literature, including engineering charts and diagrams, provided by the technology providers.

The committee recognizes and appreciates the extensive support of the Army ACWA team and its interactions with stakeholders and the Dialogue group, particularly the four members of the Dialogue known as the Citizens Advisory Technical Team (CATT). Members of the CATT attended all open meetings of the committee and shared information and their views with us.

The committee also appreciates the openness and cordiality of the representatives of the technology providers. They and the Army provided us with early drafts of their test reports and other documentation to facilitate the development of this report.

A study like this always requires extensive logistic support, and we are all indebted to the NRC staff for their assistance. I would like to acknowledge particularly the close working relationship I had with the NRC study director for this study, Dr. Patricia P. Paulette. Working as a team in leading this study, she and I spoke on the phone daily and e-mailed each other incessantly. Invaluable contributions were also made by Harrison T. Pannella, who took extensive notes at all of our meetings, edited draft text for the report, and provided suggestions for organizing the report. In addition, Jacqueline Johnson and Gwen Roby provided the logistic support that freed us to concentrate on our task. Assistance was also provided by James C. Myska. The report was edited by Carol R. Arenberg, Division on Engineering and Physical Sciences. I am also indebted to colleagues in the Chemistry Department at the University of Southern California, who willingly substituted for me in my teaching duties while I traveled on behalf of this study.

Robert A. Beaudet, *Chair*
Committee on Review and Evaluation of
Alternative Technologies for Demilitarization of
Assembled Chemical Weapons: Phase II

Acknowledgments

This report has been reviewed in draft form by individuals chosen for their diverse perspectives and technical expertise, in accordance with procedures approved by the NRC's Report Review Committee. The purpose of this independent review is to provide candid and critical comments that will assist the institution in making its published report as sound as possible and to ensure that the report meets institutional standards for objectivity, evidence, and responsiveness to the study charge. The review comments and draft manuscript remain confidential to protect the integrity of the deliberative process. We wish to thank the following individuals for their review of this report:

Milton Beychok, independent consultant
Digby McDonald, Pennsylvania State University
Alvin Mushkatel, Arizona State University
Kirk Newman, Naval Surface Warfare Center

Robert Olson, independent consultant
George Parshall, Chemical Science (retired)
Carl Peterson, Massachusetts Institute of Technology
Janice Phillips, Centocor

Although the reviewers listed above have provided many constructive comments and suggestions, they were not asked to endorse the conclusions or recommendations, nor did they see the final draft of the report before its release. The review of this report was overseen by Hyla S. Napadensky (NAE), Napadensky Energetics, Inc. (retired), appointed by the NRC's Report Review Committee, who was responsible for making certain that an independent examination of this report was carried out in accordance with institutional procedures and that all review comments were carefully considered. Responsibility for the final content of this report rests entirely with the authoring committee and the institution.

Contents

EXECUTIVE SUMMARY 1

1 INTRODUCTION 9
Background, 9
Description of the Pueblo Stockpile, 12
    Agents, 12
    Weapon Types, 14
Role of the National Research Council, 14
Statement of Task, 14
Scope of This Report, 15
Organization of This Report, 15

2 HYDROLYSIS TESTS OF ENERGETIC MATERIALS 17
Current Practices for the Disposal of Energetic Materials, 17
Caustic Hydrolysis of Energetic Materials, 18
Overview of the Test Program, 19
    Testing at the Holston Army Ammunition Plant, 19
    Bench-scale Tests at Los Alamos National Laboratory, 20
    Bench-scale Tests at the Pantex Plant, 21
    Bench-scale Tests at the Naval Surface Warfare Center, 21
    Hydrolysate Production at the Radford Army Ammunition Plant, 21
Program Status, 21
    Results of Tests at the Holston Army Ammunition Plant, 21
    Results of Tests at Los Alamos National Laboratory, 22
    Results of Tests at the Naval Surface Warfare Center, 22
    Analysis of an Incident at Radford Army Ammunition Plant, 22
Summary Assessment, 24
    Previous Findings and Recommendations of the ACW I Committee, 24
    New Findings and Recommendations, 25

3 GENERAL ATOMICS TECHNOLOGY PACKAGE　　27
   Description of the Process, 27
       Disassembly of Munitions (Steps 1 to 4), 27
       Hydrolysis of Energetic Materials (Steps 5 and 6), 27
       Separation of Agent from Munition Bodies and Agent Hydrolysis (Steps 7 to 10), 30
       Treatment of Agent Hydrolysate by Supercritical Water Oxidation (Step 11), 31
       Processing and Treatment of Dunnage and Energetics Hydrolysate (Steps 12 to 15), 32
       Water Recovery and Salt Disposal (Step 16), 33
   Information Used in the Development of the Assessment, 33
       Engineering Design Package, 33
       Engineering Design Studies Tests, 33
   Assessment of Process Component Design, 38
       Disassembly of Munitions (Steps 1 to 6), 38
       Separation of Agent from Munition Bodies and Agent Hydrolysis (Step 7), 39
       Agent Hydrolysis and Metal Parts Treatment (Steps 8 to 10), 40
       Treatment of Hydrolysates and Dunnage by Supercritical Water Oxidation
           (Steps 11 and 15), 40
       Processing and Treatment of Dunnage and Energetics Hydrolysate (Steps 12 to 16), 41
   Assessment of Integration Issues, 42
       Component Integration, 42
       Process Operability, 42
       Monitoring and Control Strategy, 42
       Maintenance Issues, 43
       Process Safety, 43
       Worker Health and Safety, 44
       Public Safety, 44
       Human Health and the Environment, 44
   Assessment of Overarching Technical Issues, 45
       Overall Engineering Design Package, 45
       Steps Required Before Implementation, 45
       Previous Findings and Recommendations, 46
       New Findings and Recommendations, 48

4 PARSONS/HONEYWELL TECHNOLOGY PACKAGE　　49
   Description of the Process, 49
       Introduction and Overview, 49
       Disassembly of Munitions and Removal of Agent and Energetics, 49
       Hydrolysis of Agent and Energetics, 52
       Biological Treatment, 54
       Metal Parts Treaters, 55
       Continuous Steam Treater for Dunnage, 56
       Treatment of Off-gases and Disposal of Wastes, 56
       Changes to Process, 58
   Information Used in the Development of the Assessment, 58
       Engineering Design-Related Documents, 58
       Engineering Design Studies Tests, 59
   Assessment of Process Component Design, 59
       Disassembly of Munitions and Removal of Agent and Energetics, 59
       Hydrolysis of Agent, 60
       Hydrolysis of Energetics, 60
       Biological Treatment, 61
       Metal Parts Treatment, 62
       Treatment of Dunnage in the Continuous Steam Treater, 62
       Off-gas Treatment and Disposal of Wastes, 63

Assessment of Integration Issues, 63
  Component Integration, 63
  Process Operability, 63
  Monitoring and Control Strategy, 64
  Process Safety, 64
  Worker Health and Safety, 64
  Public Safety, 65
  Human Health and the Environment, 66
Assessment of Overarching Technical Issues, 67
  Steps Required Before Implementation, 67
  Previous Findings and Recommendations, 67
  New Findings and Recommendations, 68

5 GENERAL FINDINGS AND RECOMMENDATIONS   71
  Engineering Design Studies, 71
  Update on General Findings and Recommendations of the ACW I Committee, 73
    General Findings from the 1999 Inital ACW I Committee Report, 73
    General Recommendations from the 1999 Initial ACW I Committee Report, 75
    General Findings from the 2000 Supplemental ACW I Committee Report, 75

REFERENCES   76

APPENDIXES

A   Description of Munitions in the Pueblo Chemical Depot Stockpile   81

B   SCWO Reliability and Maintenance (RAM) Log for 500-Hour HD Hydrolysate Run   83

C   Committee Meetings and Site Visits   88

D   Biographical Sketches of Committee Members   93

Figures and Tables

FIGURES

ES-1 Simplified block diagram of GATS process components, 4
ES-2 Parsons/Honeywell WHEAT block flow diagram, 5

3-1 Simplified schematic flow diagram for GATS, 28
3-2 Simplified block diagram of GATS process components, 29

4-1 Parsons/Honeywell WHEAT block flow diagram, 50
4-2 Agent hydrolysis process, 53
4-3 Energetics hydrolysis process, 54
4-4 Biotreatment process, 55
4-5 Off-gas treatment systems, 57

A-1 105-mm howitzer projectile, 81
A-2 155-mm howitzer projectile, 82
A-3 4.2-inch mortar cartridge, 82

TABLES

1-1 Descriptions of the Seven Technology Packages That Passed the Go/No-Go Evaluation, 11
1-2 Munitions Containing HD and HT in the Pueblo Chemical Depot Stockpile, 13
1-3 Physical Properties of Mustard Agents at Pueblo Chemical Depot, 13
1-4 Original Nominal Composition of HD Mustard, 14
1-5 Original Composition of HT Mustard, 14

2-1 Nominal Composition of Energetic Materials Used in Chemical Munitions, 17

3-1 Design Parameters for GATS ERH and PRH, 29
3-2 Key Design Parameters for GATS Cryofracture Systems (Two Trains), 30
3-3 Key Design Parameters for the GATS Projectile Agent Hydrolysis System, 31
3-4 Equipment Sizes for the Full-scale SCWO System, 31
3-5 Design Parameters for the GATS DSHS, 32
3-6 Feeds and Duration of Planned SCWO Tests, 35
3-7 Corrosion of Titanium Liners During GATS EDS Work-up Tests, 36

4-1 Changes to the Parsons/Honeywell Process Since Demonstration I, 58

B-1 SCWO Reliability and Maintenance (RAM) Log for 500-Hour HD Hydrolysate Run, 84

Acronyms and Abbreviations

ACAMS	automatic continuous air monitoring system
ACWA	Assembled Chemical Weapons Assessment
ARDEC	Armament Research, Development and Engineering Center
BPCS	basic process control system
BWM	burster washout machine
CAMDS	Chemical Agent Munitions Disposal System
CATOX	catalytic oxidation
CATT	Citizens Advisory Technical Team
CSDP	Chemical Stockpile Disposal Program
CST	continuous steam treater
CWC	Chemical Weapons Convention
DOD	U.S. Department of Defense
DPE	demilitarization protective ensemble
DRE	destruction and removal efficiency
DSHS	dunnage-shredder hydrolysis system
ECR	explosion containment room
EDP	engineering design package
EDS	engineering design study
EPA	Environmental Protection Agency
ERD	energetics rotary deactivator
ERH	energetics rotary hydrolyzer
ESS	emergency shutdown system
EST	engineering-scale test
GATS	General Atomics Total Solution (technology package)
GB	a nerve agent
H	undistilled mustard agent
HAAP	Holston Army Ammunition Plant
HD	distilled mustard agent
HDC	heated discharge conveyor
HEPA	high-efficiency particulate air
HMX	cyclotetramethylene-tetranitramine (an energetic material)
HRA	health risk assessment
HT	a type of mustard agent containing mustard-T
HVAC	heating, ventilating, and air conditioning
ICB	immobilized-cell bioreactor
IITRI	Illinois Institute of Technology Research Institute
JACADS	Johnston Atoll Chemical Agent Disposal System
kW	kilowatt
LANL	Los Alamos National Laboratory
M	molar concentration
MAV	modified ammunition van
MDB	munitions demilitarization building
MDM	munitions demilitarization machine
MPT	metal parts treater
MSB	munitions storage building
NEPA	National Environmental Policy Act
NRC	National Research Council
NSWC	Naval Surface Warfare Center
OB/OD	open burn/open detonation
PHA	preliminary hazards analysis
PLC	programmable logic control/controller
PMACWA	Program Manager for Assembled Chemical Weapons Assessment
PMCD	Program Manger for Chemical Demilitarization
PMD	projectile mortar demilitarization (machine)
PRH	projectile rotary hydrolyzer
PRR	propellant removal room

psi	pounds per square inch	WHEAT	water hydrolysis of explosives and agent technology
psig	pounds per square inch gauge	WMDM	WHEAT multipurpose demilitarization machine
QRA	quantitative risk assessment	WPMD	WHEAT projectile/mortar disassembly (machine)

R3   resource reclamation and recycling (process)
R&D   research and development
RAAP   Radford Army Ammunition Plant
RCRA   Resource Conservation and Recovery Act
RDX   an energetic material
RFP   Request for Proposals
ROD   Record of Decision
RWM   rotary washout machine

scfm   standard cubic feet per minute
SCWO   supercritical water oxidation

T   a mustard ether
TACOM   Tank-Automotive and Armaments Command
TCLP   toxicity characteristic leaching procedure
TNT   trinitrotoluene (an energetic material)
TOC   total organic carbon

UPA   unpack area

VX   a nerve agent

3X   At the 3X decontamination level, solids are decontaminated to the point that agent concentration in the headspace above the encapsulated solid does not exceed the health-based, eight-hour, time-weighted average limit for worker exposure. The level for mustard agent is 3.0 µg per cubic meter in air. Materials classified as 3X may be handled by qualified plant workers using appropriate procedures but are not releasable to the environment or for general public reuse. In specific cases in which approval has been granted, a 3X material may be shipped to an approved hazardous waste treatment facility for disposal in a landfill or for further treatment.

5X   Treatment of solids to a 5X decontamination level is accomplished by holding a material at 1,000°F for 15 minutes. This treatment results in completely decontaminated material that can be released for general use or sold (e.g., as scrap metal) to the general public in accordance with applicable federal, state, and local regulations.

Executive Summary

The Program Manager for Assembled Chemical Weapons Assessment (PMACWA) of the Department of Defense (DOD) requested the National Research Council (NRC) to assess the engineering design studies (EDSs) developed by Parsons/Honeywell and General Atomics for a chemical demilitarization facility to completely dispose of the assembled chemical weapons at the Pueblo Chemical Depot in Pueblo, Colorado. To accomplish the task, the NRC formed the Committee on Review and Evaluation of Alternative Technologies for Demilitarization of Assembled Chemical Weapons: Phase II (ACW II Committee). This report presents the results of the committee's scientific and technical assessment, which will assist the Office of the Secretary of Defense in selecting the technology package for destroying the chemical munitions at Pueblo. The Record of Decision (ROD) for selecting the technology package is expected in the second half of 2001.

The committee evaluated the engineering design packages proposed by the technology providers and the associated experimental studies that were performed to validate unproven unit operations. A significant part of the testing program involved expanding the technology base for the hydrolysis of energetic materials associated with assembled weapons. This process was a concern expressed by the Committee on Review and Evaluation of Alternative Technologies for Demilitarization of Assembled Chemical Weapons (ACW I Committee) in its original report in 1999 (NRC, 1999). The present study took place as the experimental studies were in progress. In some cases, tests for some of the supporting unit operations were not completed in time for the committee to incorporate results into its evaluation. In those cases, the committee identified and discussed potential problem areas in these operations. Based on its expertise and its aggressive data-gathering activities, the committee was able to conduct a comprehensive review of the test data that had been completed for the overall system design.

This executive summary is divided into four sections. The first section provides historical background for the DOD's program for chemical demilitarization and the NRC's involvement. The next section gives the statement of task for the ACW II Committee's studies. The third section briefly describes the technologies and test programs assessed in this report, and the final section presents the committee's general findings. Detailed findings and recommendations found in the chapters relating to the individual technologies are not repeated here, but they may be found at the end of each chapter.

HISTORICAL BACKGROUND

The U.S. Army is in the process of destroying the U.S. stockpile of aging chemical weapons, which is stored at eight locations in the continental United States and on Johnston Atoll in the Pacific Ocean. The deadline for completing the destruction of these weapons, as specified by the Chemical Weapons Convention (CWC) international treaty, is April 29, 2007. Originally, the Army selected incineration as the preferred baseline destruction technology, and it currently operates two incineration facilities—one on Johnston Atoll and one at the Deseret Chemical Depot near Tooele, Utah. The Johnston Atoll Chemical Agent Disposal System completed destruction of the stockpile on Johnston Island in late 2000, and plans for closure of the facility are under way.[1] Similar baseline incineration system facilities were planned for all of the remaining storage sites. However, incineration has met with public and political opposition. In response to this opposition, neutralization processes (based on the hydrolysis of chemical agent using either water or sodium hydroxide solution) have been developed to destroy the chemical agents stored in bulk containers at Aberdeen, Maryland, and Newport, Indiana. For the remaining sites, where

---
[1]The stockpile on Johnston Island comprised 2,031 tons, or 6.4 percent, of the original 31,496 tons of chemical nerve and blister (mustard) agents in the U.S. stockpile.

munitions containing both chemical agent and energetic materials (i.e., assembled chemical weapons) are stored, incineration is still the planned approach for destruction. In late 1996, however, Congress enacted Public Law 104-201, which instructed DOD to "conduct an assessment of the chemical demilitarization program for destruction of assembled chemical munitions and of the alternative demilitarization technologies and processes (other than incineration) that could be used for the destruction of the lethal chemical agents that are associated with these munitions."

Another law, Public Law 104-208, required a new program manager (the Program Manager for Assembled Chemical Weapons Assessment) to "identify and demonstrate not less than two alternatives to the baseline incineration process for the demilitarization of assembled chemical munitions." In addition, the law prohibited any obligation of funds for the construction of incineration facilities at two storage sites—Lexington/Blue Grass, Kentucky, and Pueblo, Colorado—until the demonstrations were completed and an assessment of the results had been submitted to Congress by DOD.

As a result of Public Laws 104-201 and 104-208, DOD created the Assembled Chemical Weapons Assessment (ACWA) program. To ensure public involvement in the program, the PMACWA enlisted the Keystone Center—a nonprofit, neutral facilitation organization—to convene a diverse group of interested stakeholders, called the Dialogue on ACWA (or, simply, the Dialogue), who would be intimately involved in all phases of the program. The 35 members of the Dialogue include representatives of the affected communities, national citizen groups such as the Sierra Club, state regulatory agencies, affected Native American tribes, the Environmental Protection Agency (EPA), and DOD.

The PMACWA established an elaborate program for evaluating and selecting technologies that would be appropriate for destroying the stockpile at Pueblo Chemical Depot and Blue Grass Chemical Depot. The selection process is described in detail in the 1999 NRC report *Review and Evaluation of Alternative Technologies for the Demilitarization of Assembled Chemical Weapons*. Six technology packages were originally considered for the demonstration tests. Three of these technologies underwent demonstration testing in the first round (Demonstration I) and two technology packages survived as candidates for the destruction of chemical weapons at the Pueblo Chemical Depot: those of General Atomics and Parsons/Honeywell. In Public Law 105-261 (1999), Congress mandated as follows: "The program manager for the Assembled Chemical Weapons Assessment shall continue to manage the development and testing (including demonstration and pilot-scale testing) of technologies for the destruction of lethal chemical munitions that are potential or demonstrated alternatives to the baseline incineration program." It also directed that the Army continue its coordination with the NRC. The PMACWA subsequently initiated EDSs for the two technologies that successfully completed demonstration testing. The purpose of this EDS phase is to (1) support the development of a Request for Proposals (RFP) for a pilot facility; (2) support the certification decision of the Under Secretary of Defense for Acquisition and Technology, as directed by Public Law 105-261; and (3) support documentation required for the National Environmental Policy Act (NEPA) and the data required for a permit under the Resource Conservation and Recovery Act (RCRA). Each EDS comprises two parts: an engineering design package (EDP) and the results of experimental studies conducted to generate required data that were not obtained during the demonstration test phase.

In response to Public Law 104-201, which required that DOD coordinate its efforts with the NRC in assessing alternatives to incineration, PMACWA asked the NRC to evaluate each of the seven technologies that had passed DOD's initial screening. The ACW I Committee published its report in August 1999. That report found that the primary treatment processes could decompose the chemical agents with destruction efficiencies of 99.9999. However, major concerns for each technology package remained, including the adequacy of secondary treatment of agent hydrolysates and the primary and secondary treatment of energetic materials contained in the chemical weapons. A supplemental report, requested by the PMACWA to evaluate the actual demonstration tests for the three technologies that were considered to warrant further investigation, was published in February 2000. Two of the technologies, those of General Atomics and Parsons/Honeywell, were considered ready to proceed to an engineering design phase. Upon completion of the supplemental report, the ACW I Committee was dissolved. Subsequently, under the continuing mandate from Congress, the PMACWA requested that the NRC form a second committee (the ACW II Committee) to evaluate the EDPs and related tests for the engineering design studies for the Pueblo and Blue Grass Depots and to examine and evaluate the Demonstration II tests of three additional technologies.

STATEMENT OF TASK

The statement of task for the NRC ACW II Committee is shown below. The present report is the committee's response to Task 2 and will be produced in time to contribute to the ROD by the Office of the Secretary of Defense on a technology selection for the Pueblo site. The latter will occur following satisfaction of NEPA requirements.

At the request of the DoD's Program Manager for Assembled Chemical Weapons Assessment (PMACWA), the NRC Committee on Review and Evaluation of Alternative Technologies for Demilitarization of Assembled Chemical Weapons will provide independent scientific and technical assessment of the Assembled Chemical Weapons Assessment (ACWA) program. This effort will be divided into three tasks. In each case, the NRC was asked to perform a techni-

Task 1

To accomplish the first task, the NRC will review and evaluate the results of demonstrations for three alternative technologies for destruction of assembled chemical weapons located at U.S. chemical weapons storage sites. The alternative technologies to undergo demonstration testing are: the AEA Technologies electrochemical oxidation technology, the Teledyne Commodore solvated electron technology, and the Foster Wheeler and Eco Logic transpiring wall supercritical water oxidation and gas phase chemical reduction technology. The demonstrations will be performed in the June through September 2000 timeframe. Based on receipt of the appropriate information, including: (a) the PMACWA-approved Demonstration Study Plans, (b) the demonstration test reports produced by the ACWA technology providers and the associated required responses of the providers to questions from the PMACWA, and (c) the PMACWA's demonstration testing results database, the committee will:

- perform an in-depth review of the data, analyses, and results of the unit operation demonstration tests contained in the above and update as necessary the 1999 NRC report, *Review and Evaluation of Alternative Technologies for Demilitarization of Assembled Chemical Weapons* (the ACW report)
- determine if any of the AEA Technologies, Teledyne Commodore, and Foster Wheeler/Eco Logic technologies have reached a technology readiness level sufficient to proceed with implementation of a pilot-scale program
- produce a report for delivery to the PMACWA by July 2001 provided the demonstration test reports are made available by November 2000. (An NRC report delivered in March 2000 covered the initial three technologies selected for demonstration phase testing.)

Task 2

For the second task, the NRC will assess the ACWA Engineering Design Study (EDS) phase in which General Atomics and Parsons/Honeywell (formerly Parsons/Allied Signal) will conduct test programs to gather the information required for a final engineering design package representing a chemical demilitarization facility at the Pueblo, Colorado stockpile site. The testing will be completed by September 1, 2000. Based on receipt of the appropriate information, including: (a) the PMACWA-approved EDS Plans, (b) the EDS test reports produced by General Atomics and Parsons/Honeywell, (c) PMACWA's EDS testing database, and (d) the vendor-supplied engineering design packages, the committee will:

- perform an in-depth review of the data, analyses, and results of the EDS tests
- assess process component designs, integration issues, and overarching technical issues pertaining to the General Atomics and the Parsons/Honeywell engineering design packages for a chemical demilitarization facility design for disposing of mustard-only munitions
- produce a report for delivery to the PMACWA by March 2001 provided the engineering design packages are received by October 2000.

Task 3

For the third task, the NRC will assess the ACWA EDS phase in which General Atomics will conduct test programs to gather the information required for a final engineering design package representing a chemical demilitarization facility at the Lexington/Blue Grass, Kentucky stockpile site. The testing will be completed by December 31, 2000. Based on receipt of the appropriate information, including: (a) the PMACWA-approved EDS Plans, (b) the EDS test reports produced by General Atomics, (c) PMACWA's EDS testing database, and (d) the vendor-supplied engineering design package, the committee will:

- perform an in-depth review of the data, analyses, and results of the EDS tests
- assess process component designs, integration issues, and overarching technical issues pertaining to the General Atomics engineering design package for a chemical demilitarization facility design for disposing of both nerve and mustard munitions
- produce a report for delivery to the PMACWA by September 2001 provided the engineering design package is received by January 2001.

DESCRIPTION OF THE TECHNOLOGY PACKAGES

The assembled chemical weapons at Pueblo contain only mustard agent and energetic materials. The operations required for their destruction include (1) unpacking and disassembling the weapons, (2) separation of agents, energetics, and metal parts, (3) destruction of agent and energetic hydrolysates, (4) decontamination of the metal parts, (5) destruction of the dunnage, and (6) treatment and disposal of all associated solid, liquid, and gaseous by-products.

For both the General Atomics and the Parsons/Honeywell design packages, the primary treatment to destroy the agent and the energetic materials is hydrolysis. However, the hydrolysis products (hydrolysates) must be further treated before the final products can be properly disposed of. For this secondary step, General Atomics proposes to use supercritical water oxidation (SCWO) and Parsons/Honeywell proposes to use biotreatment via immobilized cell bioreactors (ICBs).

Both technology packages consist of multiple unit operations that work in sequence or concurrently to carry out all aspects of chemical weapons destruction. Both processes are designed to treat agent, energetic materials, metal parts (including munitions bodies), dunnage (e.g., wooden pallets and packing boxes used to store munitions), and nonprocess waste (e.g., plastic demilitarization protective ensemble (DPE) suits; the carbon from DPE suit filters and plant heating, ventilating, and air conditioning (HVAC) filters; and miscellaneous plant wastes). Each EDP includes engineering drawings and documentation, a preliminary hazards analysis, and costs and schedule for the technology to be implemented at the Pueblo Chemical Depot. Short descriptions are given below. More detailed descriptions of the unit operations for each technology are given in Chapters 3 and 4.

Figure ES-1 is a block diagram of the General Atomics technology process, which uses the acronym GATS (General Atomics total solution). The following major operations are included:

- A modified baseline disassembly process is used; however, cryofracture is used to open the projectile bodies to access the agent. The bodies are cooled to liquid nitrogen temperature and fractured. Then the metal parts are separated from the agent.
- Agents and energetics are hydrolyzed in batch reactors to form hydrolysates.
- Fuzes are digested in an energetics rotary hydrolyzer with caustic.
- Munition bodies are decontaminated to a 5X condition by using an electrically heated discharge conveyor.
- The dunnage is shredded and slurried.
- All the resulting hydrolysates and the slurried dunnage are further treated with SCWO to produce environmentally benign products.
- System off-gases are processed through carbon filters.

The unit operations tested during the EDS phase are the dunnage shredder hydrolysis system (DSHS), the energetic rotary hydrolyzer (ERH), and the SCWO reactor. The testing of the SCWO reactor had not been completed when this report was prepared.

The Parsons/Honeywell technology team uses the acronym WHEAT (water hydrolysis of explosives and agent technology) to denote its technology package for the demilitarization of assembled chemical weapons. The process is described in Figure ES-2. It consists of the following main operations:

- The Army's baseline disassembly process, with modifications, is used to separate agent, energetics, and metal parts.
- The solid heel or sludge that remains inside the munitions casing is washed out in the projectile rotary washout machine (RWM) using recirculated wash water through high-pressure water jets.
- Bursters from the mortars and projectiles are fed into the burster washout machine (BWM) by a pick-and-place machine and processed in the BWMs to wash out all explosives.

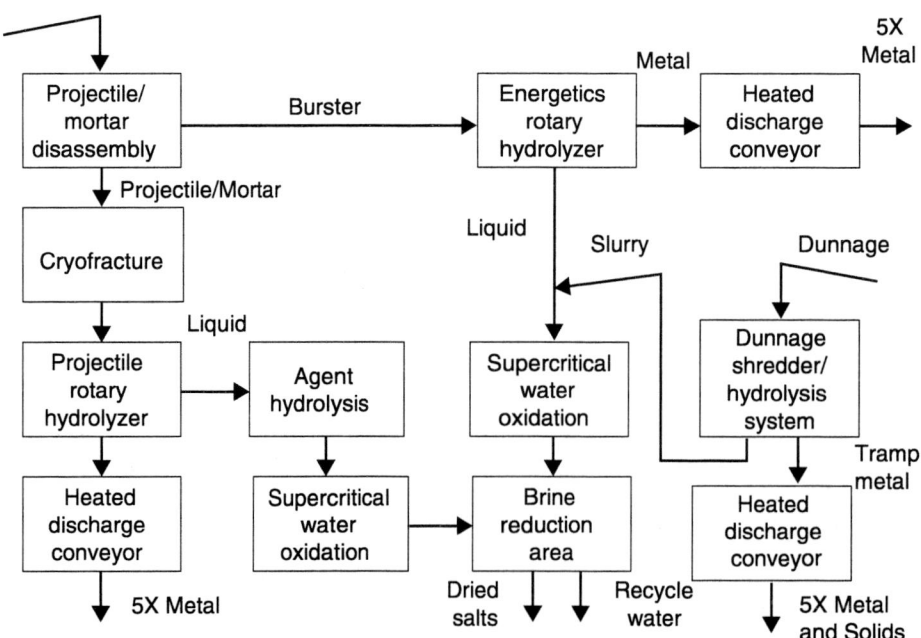

FIGURE ES-1 Simplified block diagram of GATS process components. Source: Adapted from General Atomics, 2000a.

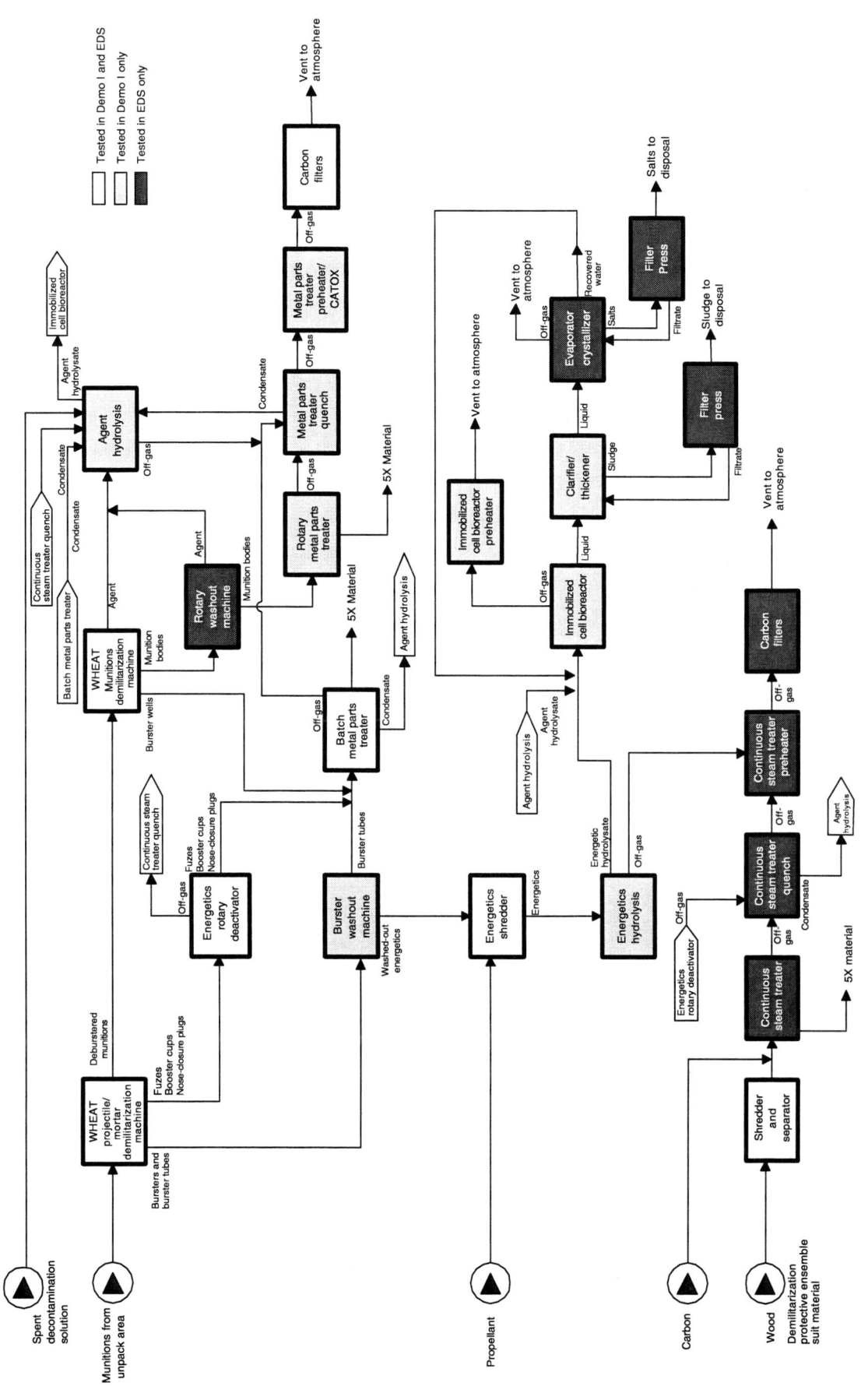

FIGURE ES-2 Parsons/Honeywell WHEAT block flow diagram. Source: Adapted from Parsons, 2000a.

- The energetics rotary deactivator (ERD) receives fuzes, booster cups, and miscellaneous parts, and it heats them until they are deflagrated.
- Agents and energetics are hydrolyzed in batch reactors to form hydrolysates.
- Agent and energetics hydrolysates are diluted with water, mixed with inorganic nutrients, and fed to the ICBs, which contain aerobic microorganisms that will consume most of the organic content of the hydrolysates.
- Biological processing, followed by evaporation/crystallization, converts the hydrolysis products to liquids or solids acceptable for discharge to the environment or liquids acceptable for recycling. Biological treatment is done in the ICBs.
- Metal parts are all treated either in the batch metal parts treater (batch MPT) or the rotary metal parts treater (rotary MPT) to decontaminate metal parts to 5X.
- Dunnage is heat treated in the continuous steam treater (CST) to decontaminate it to 5X.
- Gas discharges from the plant are passed through catalytic oxidation (CATOX) units. Some of the gas streams are also passed through activated carbon filters.

The ICB, the CST, the CATOX unit, and the projectile washout system were tested during EDS. However, the CST and the projectile washout operations were not finished at the time this report was prepared.

The committee formed two working groups to perform in-depth evaluations of each EDP. As part of their efforts, the groups visited the EDS test sites at Aberdeen Proving Ground, Maryland; Dugway Proving Ground, Utah; and Deseret Chemical Depot, Utah. Committee members also attended PMACWA status-review meetings, which were held periodically, and a review meeting at Parsons/Honeywell in Pasadena, California, where both Parsons/Honeywell and General Atomics personnel described their EDPs and the results of ongoing tests. The technology providers and PMACWA staff kindly provided draft copies of reports as they were generated. The final EDPs were released in October 2000.

In evaluating the general efficacy of the design plans for a chemical demilitarization facility suited to the Pueblo Chemical Depot and the readiness of each technology to go forward to the next level of pilot plant testing, the committee relied upon its knowledge of the proposed systems, available test results, aggressive data collection activities, and thorough review of the engineering design plans.

GENERAL FINDINGS

General findings on the EDS phase of the ACWA program for the two technology packages evaluated in this report appear below. The general findings must be considered with acknowledgment of the fact that some ACWA EDS testing was not completed in time for the committee to obtain final test results and that some process steps remain to be demonstrated on a pilot scale. Specific findings and recommendations for each technology package, as well as the PMACWA-sponsored investigations on hydrolysis of energetic materials, appear in the body of the report. The energetics hydrolysis test program is progressing at a pace satisfactory to meet the engineering requirements for construction of a disposal facility at Pueblo Chemical Depot. Issues surrounding the hydrolysis of neat tetryl, optimum granulation sizes, more complete characterization of hydrolysis products from aromatic nitro compounds, and optimum process control strategies for full-scale operations are yet to be investigated.

General Finding (Pueblo) 1. Based on the results of the demonstration tests, the engineering design package, and available data, the committee believes that the Parsons/Honeywell WHEAT technology package can provide an effective and safe means of destruction for the assembled chemical weapons stored at the Pueblo Chemical Depot. However, some of the process steps remain to be demonstrated.

The Parsons/Honeywell technology process provides effective means to:

- disassemble munitions by a modified baseline disassembly process that removes the agent from the projectile bodies by washout
- destroy chemical agent HD to a 99.9999 percent destruction and removal efficiency (DRE) by hydrolysis
- destroy fuzes with the energetics rotary deactivator
- destroy energetic materials to a 99.999 percent DRE by hydrolysis in 15 weight percent hot caustic solution, provided that the following safeguards are observed:
  —different energetic materials are not processed together
  —precautions are taken to ensure that all emulsified TNT is completely destroyed
- control the very large volumes of off-gases emitted from the biotreatment plant through a CATOX unit

However, the committee notes that the effectiveness of some process steps, including removal of energetics from munitions, has not been tested during the EDS. Treatment of metal parts, dunnage, and DPE suit material remains to be demonstrated. No tests are currently planned to demonstrate the efficacy of the burster washout and energetic materials size-reduction steps. The projectile washout system is currently being tested. Other remaining munition disassembly operations are very similar to those used in the baseline system and have therefore been proven. The energetics rotary deactivator concept appears workable but has not been demonstrated at the pilot scale. Energetics hydrolysis is relatively immature, but current testing at Holston AAP has the capability to resolve many, but not all, of these issues (see Chapter 2).

The testing of the continuous steam treater for dunnage and the projectile washout system will not be complete until October 2001. Dioxins and furans are present in the off-gas from the CATOX units on the bioreactors but are below levels of regulatory concern. The batch metal parts treater for small metal parts is being tested, and preliminary data are encouraging. The carousel fixture for the rotary metal parts treater for large metal parts has not been demonstrated. The use of catalytic oxidizers for various streams is currently being tested, but sufficient test data have not been provided to the committee. Because the honeycomb structure of the CATOX unit is susceptible to plugging, proper design must be employed to prevent particulates from entering the catalyst structure.

General Finding (Pueblo) 2. Based on the results of the demonstration tests, the engineering design package, and available data, the committee believes that many aspects of the General Atomics technology package can be effective and safe for the destruction of assembled chemical weapons at the Pueblo Chemical Depot. However, to achieve prolonged operability of the SCWO system as designed will require extensive maintenance. In addition, the SCWO processing of dunnage slurried in energetics hydrolysate, which constitutes the vast majority of the feedstock to be processed, remains unproven. The viability of the General Atomics technology package will depend on acceptable operability of the SCWO system.

The General Atomics technology process provides effective means to:

- disassemble munitions by using a modified baseline disassembly process for munitions and removal of the agent from the projectile bodies by cryofracture
- destroy chemical agent HD to a 99.9999 percent DRE by hydrolysis
- destroy fuzes with the energetics rotary hydrolyzer
- destroy energetic materials to a 99.999 percent DRE by hydrolysis in 15 weight percent hot caustic solution, provided that the following safeguards are observed:
  — different energetic materials are not processed together
  — precautions are taken to ensure that all emulsified TNT is completely destroyed
- provide effective 5X-level decontamination for munition bodies through the use of an electrically heated discharge conveyor
- readily control the very low volumes of off-gases produced through activated carbon adsorption systems

For dunnage, the materials are shredded and reduced in size to 1.0 mm. The slurry is then fed into the SCWO reactors to destroy all the dunnage.

However, the committee has serious concerns about the SCWO system that is used to process the hydrolysates and the slurried dunnage. At the time this report was prepared, not all of the long-term processing tests had been completed. On the basis of results to date, the committee has concerns about the ability of the SCWO reactor to operate continuously for adequate lengths of time. An additional concern is the ability of the size reduction system to remove 100 percent of the tramp metal that comes with the dunnage. If the tramp metal is not removed from the dunnage, the committee believes it will clog the injectors of the SCWO system and further reduce the system's online availability.

The SCWO tests that have been performed to date, especially those involving chlorinated organic compounds such as HD hydrolysate, have consistently encountered severe corrosion of the reactor material or plugging of the reactor with salts. General Atomics proposes to solve the problem of plugging by periodically (every 22 hours of operation) reducing the pressure of the reactor to slightly below the critical point of water and flushing with clean water for 2 hours to remove the accumulated salts. The technology provider proposes to deal with the corrosion problem by inserting into the SCWO reactor a sacrificial titanium liner and shutting down at approximately every 140 hours of operation to open the reactor and replace or reverse the liner.[2] In the committee's opinion, the flushing step does not pose an unreasonable operating requirement; however, it considers the need for a liner replacement at six-day intervals to be excessively disruptive and not in keeping with sound principles of effective operation. In the full-scale system, liner replacement will require the following steps:

1. Cooling down and depressurizing the reactor,
2. Unbolting and removing an approximately 16-inch-diameter, several-inch-thick pressure head from the top of the reactor,
3. Withdrawing the 12.5-inch-diameter, 19-foot-long titanium liner from the tubular SCWO reactor,
4. Reinserting the same liner reversed end to end or a new liner,
5. Setting the pressure gasket back into place and reattaching the gasket coolant lines,
6. Resetting and bolting the pressure head onto the reactor,
7. Pressure testing the SCWO reactor to assure proper head seating and sealing, and
8. Restarting the heat-up of the system and restarting the waste feed.

---

[2]The corrosion is restricted to the top part of the liner so each liner can be used twice by opening the reactor and reinstalling it in the reactor with the uncorroded lower part up.

This appears to the committee to be a very time-consuming procedure. The experience of a number of committee members has been that large pieces of high-pressure equipment are very difficult and time consuming to seal. Tests have only been conducted with reactors 2 inches to 4 inches in diameter. The time required for this procedure at the far larger size of the full-scale SCWO unit is highly uncertain.

General Atomics proposes to build duplicate SCWO reactors so that one is operating while the second is being serviced; however, the committee has reservations about whether this level of redundancy is adequate to maintain the proposed operating schedule.

General Finding (Pueblo) 3. As the ACW I Committee observed, the unit operations in both the General Atomics GATS and the Parsons/Honeywell WHEAT technology packages have never been operated as total integrated processes. As a consequence, a prolonged period of systemization will be necessary for both to resolve integration issues as they arise, even for apparently straightforward unit operations.

This finding continues to be valid following development of and testing for the EDS design packages for the General Atomics and Parsons/Honeywell technologies. Also, in both cases, some of the routine unit operations have not yet been designed or tested. Thus, although they appear straightforward, these unit operations could require some redesign during systemization.

General Finding (Pueblo) 4. Several of the unit operations in both the General Atomics and Parsons/Honeywell processes are intended to treat process streams that are not unique to the chemical weapons stockpile and that could potentially be treated at existing off-site facilities. These streams include agent-free energetics, dunnage, brines from water recovery, and hydrolysates. Off-site treatment would simplify the overall processes and facilitate process integration by eliminating the need for further development of these unit operations. It might also simplify design requirements to meet safety concerns.

All of the process streams that could potentially be treated off-site have compositions similar to waste streams routinely treated by commercial industrial waste treatment facilities and do not exhibit any unique toxicity. Thus, they could be transported by standard commercial conveyance to commercial facilities that are appropriately permitted to receive the waste.

1

Introduction

BACKGROUND

The United States has maintained a stockpile of highly toxic chemical warfare agents and munitions for more than half a century. These chemical agents are designed to be lethal upon exposure. Stored as components of aging weapons systems, they present a growing risk to surrounding communities.

The need to destroy the aging U.S. chemical stockpile has been a long-standing concern of government, citizens, and the military. In 1985, Public Law 99-145 mandated an "expedited" effort to dispose of one particular type of chemical munitions, the M55 rocket, which could self-ignite during storage if the stabilizer in the propellant were depleted. The mandate concerning rockets was soon expanded into the U.S. Army's Chemical Stockpile Disposal Program (CSDP), whose mission is to eliminate the entire stockpile of unitary[1] chemical weapons. The CSDP developed the current baseline system, which uses incineration to destroy the agents, energetic materials, and munition packing materials (known as dunnage). The baseline system also uses a furnace to decontaminate the residual metal parts. In 1997, after having set several intermediate goals and dates for completing the destruction of the U.S. chemical weapons stockpile, Congress ratified the President's signing of the Chemical Weapons Convention (CWC), which mandates that destruction be completed by April 29, 2007.

The CSDP currently operates two baseline incineration systems facilities—one on Johnston Atoll in the Pacific Ocean and one at the Deseret Chemical Depot near Tooele, Utah. Together, these two facilities are expected to destroy approximately one-half of the total U.S. stockpile, the remainder of which is dispersed among seven other storage sites in the continental United States.[2] Similar incineration systems were initially planned for all of these sites. However, incineration has met with strong public and political opposition. In response to this opposition, neutralization processes (i.e., processes based on the hydrolysis[3] of chemical agent in water or sodium hydroxide solution) have been developed to destroy the chemical agents stored in bulk containers at Aberdeen, Maryland, and Newport, Indiana. The construction of these facilities is under way. For the remaining sites, where explosively configured "assembled" chemical weapons are stored, incineration remains the technology planned for disposal. Construction of baseline incineration facilities is proceeding at storage sites in Anniston, Alabama; Umatilla, Oregon; and Pine Bluff, Arkansas.

In 1996, Congress enacted two laws that created and appropriated funding for a new program, the Assembled Chemical Weapons Assessment (ACWA) program. Public Law 104-201 (authorization) and Public Law 104-208 (appropriation) mandated that the Department of Defense (DOD) conduct

---

[1]Unitary chemical weapons are single chemicals loaded in munitions or stored as lethal materials. More recent binary munitions have two relatively safe chemicals loaded into separate compartments and mixed to form a lethal agent only after the munition is fired or released. The components of binary munitions, which are stockpiled in separate states, are not included in the present CSDP. However, under the Chemical Weapons Convention, they are included in the munitions that will ultimately be destroyed.

[2]The Johnston Atoll Chemical Agent Disposal System (JACADS) completed destruction of the stockpile located on Johnston Island in late 2000, and plans for closure of the facility are under way. The stockpile on Johnston Island comprised 2,031 tons, or 6.4 percent, of the original 31,496 tons of chemical nerve and blister (mustard) agents in the U.S. stockpile.

[3]Hydrolysis is a reaction of a target compound with water, an acid, or a base in which some chemical bond is broken in the target and OH^- or H^+ is inserted into the bond cleavage. The destruction of chemical agent via hydrolysis is often referred to as chemical neutralization, based on the military definition of neutralize: to render something unusable or destroyed and nonfunctional. Technically, neutralization is a chemical reaction between an acid and a base to form a salt and water. Chemical agents are neither acids nor bases, however, and the use of the term neutralization for two very different processes is somewhat confusing. Nevertheless, in the literature on chemical demilitarization, the terms neutralization and hydrolysis have been used interchangeably. Therefore, unless otherwise specified, neutralization will refer to the destruction of chemical agent via hydrolysis. The word decontamination is also used to indicate removal, destruction, or neutralization of chemical agents.

an assessment of alternative technologies to the baseline incineration process for the demilitarization of assembled chemical weapons and that at least two technologies be demonstrated. Congress included the following stipulations:

- All funds for constructing stockpile disposal facilities at Blue Grass Depot in Richmond, Kentucky, and Pueblo Chemical Depot in Pueblo, Colorado, should be frozen.
- DOD should select a program manager who was not and had never been associated with the Army's program for disposal of the stockpile by incineration.

In December 1996, DOD appointed the deputy to the commander, Soldier Biological Chemical Command, to be the Program Manager for the ACWA program (PMACWA). Public Law 104-201 also required that the PMACWA conduct the assessment "in coordination with the National Research Council (NRC)," which has a standing committee, the Committee on the Review and Evaluation of the Army Chemical Stockpile Disposal Program (the Stockpile Committee), that provides technical oversight and counsel to the Army on the CSDP, including the neutralization facilities under construction in Aberdeen, Maryland, and Newport, Indiana. The Stockpile Committee could have been asked to oversee the ACWA program as well. However, in the spirit of Public Law 104-201, the PMACWA requested that the NRC establish a separate committee to conduct an independent evaluation of the alternative technologies being considered by the ACWA program. In response, the NRC formed the Committee on Review and Evaluation of Alternative Technologies for Demilitarization of Assembled Chemical Weapons (ACW I Committee).

On July 28, 1997, after organizing a staff and establishing a program plan, the PMACWA published a Request for Proposals (RFP) for a "total solution" for the destruction of assembled chemical weapons without using incineration (U.S. Army, 1997).[4] Twelve proposals were submitted in September 1997. Of these, seven passed the threshold requirements stipulated in the RFP. These technologies are summarized in Table 1-1. On July 29, 1998, after an elaborate, multitiered selection process, three technology packages were selected for demonstration testing (Demonstration I). Detailed descriptions of the selection process and all seven technologies are available in the PMACWA's two annual reports to Congress (DOD, 1997, 1998) and in the NRC report by the ACW I Committee (NRC, 1999).

Constrained by both time and budget, the PMACWA then identified unit operations that were "most critical [and] least proven" for the three technology packages selected for the demonstration tests. These unit operations had not been previously used in the disposal of chemical munitions, nor had they been integrated into a complete system for this application. Two of the three technology packages use base hydrolysis as the primary treatment step to destroy agent and energetic materials. Because most of the uncertainties concerning these technology packages pertain to the secondary treatment of products from the primary treatment step, the PMACWA provided hydrolysates for nerve agents GB and VX and mustard agent HD, for testing unit operations.[5] Approximately 1,100 gallons of GB hydrolysate and 400 gallons of VX hydrolysate were produced at the Army's Chemical Agent Munitions Disposal System (CAMDS) experimental facility at the Deseret Chemical Depot in Utah. Approximately 4,200 gallons of HD hydrolysate were produced at the Army's Aberdeen Proving Ground in Maryland. The agent hydrolysates provided a representative feedstock for the demonstration tests and enabled characterization of the intermediate product stream for residual agent, including Schedule 2 compounds (agent precursor compounds, as defined by the international CWC).

Various types and amounts of energetic materials contained in the weapons were reacted with caustic solutions similar to those specified in the technology package proposals to produce hydrolysates for the demonstration tests. Systemization (preoperational testing) was conducted from January to March 1999, and demonstrations began in March 1999 and were completed in May 1999. The technology providers submitted their reports on the demonstration tests to the PMACWA on June 30, 1999 (Burns and Roe, 1999; General Atomics, 1999a; Parsons-AlliedSignal, 1999). The PMACWA used these reports and other information to prepare the *Supplemental Report to Congress*, submitted on September 30, 1999. The PMACWA concluded that two of

---

[4]The Program Manager for Chemical Demilitarization (PMCD) defines a total solution as one capable of demilitarizing and disposing of all components and process-related materials of a fully assembled chemical weapon (U.S. Army, 1997). These components/materials encompass, at a minimum, the following:

- chemical agents GB, VX, and/or H, HD, HT
- fuzes
- explosives
- propellant
- metal parts
- fiberglass containing polychlorinated biphenyls
- wooden and fiberboard dunnage
- protective clothing made of butyl rubber, chlorinated polymers, and silicone rubber
- various plant process wastes, including aqueous decontamination solutions, synthetic/hydrocarbon hydraulic fluids, pumps, valves, motors, and mechanical equipment, and
- carbon filter media

[5]Nerve agents are organophosphonate compounds: they contain phosphorus double-bonded to an oxygen atom and single-bonded to a carbon atom. GB is O-isopropyl methylphosphonofluoridate. VX is O-ethyl-S[2-(diisopropyl amino) ethyl]-methylphosphonothiolate. Bis(2-chloroethyl) sulfide is the proper chemical name for mustard agent HD and HT (a thickened form). Mustard gas, sulfur mustard, and yperite have also been used for this agent. The term "mustard gas" is often used, but the chemical is a liquid at ambient temperature.

TABLE 1-1 Descriptions of the Seven Technology Packages That Passed the Go/No-Go Evaluation

Technology Provider	Access to Munitions	Treatment of Agent	Treatment of Energetics	Treatment of Metal Parts	Treatment of Dunnage
AEA Technology	Modified reverse assembly (high-pressure wash, new rocket shearing).	Electrochemical oxidation using silver ions in nitric acid (SILVER II).	Treated with SILVER II process.	High-pressure acid wash; thermal treatment to 5X.[a]	Shredded and treated with SILVER II process.
ARCTECH	Modified reverse assembly.	Hydrolysis with a-HAX (humic acid and strong base, KOH).	Hydrolysis with a-HAX.	Hydrolysis with a-HAX; shipped to Rock Island Arsenal for 5X treatment.	Hydrolysis with dilute a-HAX; shipped to landfill.
Burns and Roe	Modified reverse assembly.	Plasma arc.	Plasma arc.	Melted in plasma arc.	Shredded; processed in plasma arc.
General Atomics	Modified reverse assembly; cryofracture for projectiles.	Hydrolysis; supercritical water oxidation (SCWO).	Hydrolysis; SCWO.	Hydrolysis; thermal treatment to 5X.	Shredded; destroyed in SCWO.
Lockheed Martin	Modified reverse assembly (multiple lines, compact layout, new drain and wash).	Hydrolysis; SCWO; Eco Logic gas-phase chemical reduction (GPCR).	Hydrolysis; SCWO; GPCR.	Hydrolysis; GPCR to 5X.[a]	Hydrolysis; GPCR to 5X.[a]
Parsons	Modified reverse assembly (fluid-jet cutting and energetic washout for rockets).	Hydrolysis; biotreatment.	Hydrolysis; biotreatment.	Thermal treatment to 5X.[a]	Thermal treatment to 5X.[a]
Teledyne Commodore	Fluid-jet cutting; access and drain agent; washout energetics with ammonia.	Solvated-electron process in ammonia for reduction; chemical oxidation with sodium persulfate.	Solvated-electron process in ammonia for reduction; chemical oxidation with sodium persulfate.	Wash in solvated-electron solution; oxidation to 3X;[b] ship to Rock Island Arsenal for 5X[a] treatment.	Crushed or shredded; treated in solvated electron solution; shipped to landfill.

[a]Treatment of solids to a 5X decontamination level is accomplished by holding a material at 1,000°F for 15 minutes. This treatment results in completely decontaminated material that can be released for general use or sold (e.g., as scrap metal) to the general public in accordance with applicable federal, state, and local regulations.
[b]At the 3X decontamination level, solids are decontaminated to the point that agent concentration in the headspace above the encapsulated solid does not exceed the health-based, eight-hour, time-weighted average limit for worker exposure. The level for mustard agent is 3.0 µg per cubic meter in air. Materials classified as 3X may be handled by qualified plant workers using appropriate procedures but are not releasable to the environment or for general public reuse. In specific cases in which approval has been granted, a 3X material may be shipped to an approved hazardous waste treatment facility for disposal in a landfill or for further treatment.
Source: Adapted from DOD, 1998.

the three technologies were acceptable for further development (DOD, 1999). The Burns and Roe plasma arc technology was judged to be too immature for further consideration.

In Public Law 105-261, Congress mandated as follows:

> The program manager for the Assembled Chemical Weapons Assessment shall continue to manage the development and testing (including demonstration and pilot-scale testing) of technologies for the destruction of lethal chemical munitions that are potential or demonstrated alternatives to the baseline incineration program. In performing such management, the program manager shall act independently of the program manager for Chemical Demilitarization and shall report to the Under Secretary of Defense for Acquisition and Technology.

The law also directed that the Army continue coordinating its activities with the NRC. The PMACWA initiated engineering design studies (EDSs) for the two technologies that successfully completed demonstration testing: the Parsons/

Honeywell technology[6] (hydrolysis followed by biotreatment) and the General Atomics technology (hydrolysis followed by supercritical water oxidation [SCWO]), for possible use at the Pueblo Chemical Depot in Pueblo, Colorado, and the Blue Grass Army Depot in Lexington, Kentucky. The purpose of the EDS phase was (1) to support the certification decision of the Under Secretary of Defense for Acquisition and Technology, as directed by Public Law 105-261, (2) to support the development of an RFP for a pilot facility, and (3) to support the documentation required for the National Environmental Policy Act (NEPA) and for a Resource Conservation and Recovery Act (RCRA) permit application. For each technology provider, the EDS was comprised of two parts, an engineering design package (EDP) and a set of experimental tests to generate the required additional data that was not obtained during the demonstration test phase.

In response to recommendations from the NRC, the PMACWA sponsored separate investigations to provide a basis for optimizing engineering parameters for the hydrolysis of energetic materials. Many of these investigations had not yet been completed when this report was prepared.

Thus, the following three test programs were initiated:

- The testing program sponsored by PMACWA to develop data for optimizing engineering designs for the hydrolysis of energetic materials (as discussed in Chapter 2),
- The testing program by General Atomics to develop data to support engineering design studies of its technology package (as discussed in Chapter 3), and
- The testing program by Parsons/Honeywell to develop data to support engineering design studies of its technology package (as discussed in Chapter 4).

Contracts were awarded in March 2000, and preliminary EDPs were drafted by the technology providers in June 2000. Each EDP includes drawings and documentation, a preliminary hazards analysis, and projected life-cycle costs and schedules for the technology package to be implemented at a particular site. The final EDPs were released in December 2000. Experimental tests to support the EDPs were begun in June 2000, but some had not been completed when this report was prepared. Final reports are expected to be published in mid-2001.

In 2000, Congress passed Public Law 106-79 mandating that the PMACWA "conduct evaluations of three additional alternative technologies under the ACWA program. Proceed under the same guidelines as contained in Public Law 104-208 and continue to use the Dialogue process and Citizens' Advisory Technical Team and their consultants."[7] To fulfill this mandate, the PMACWA initiated the Demonstration II program to demonstrate the three technologies not selected during the first phase.

In response to the direction of Congress, a second NRC committee, the Committee on Review and Evaluation of Alternative Technologies for Demilitarization of Assembled Chemical Weapons: Phase II (ACW II Committee), was formed in spring 2000. The new committee was asked to produce three reports: (1) an evaluation of the new demonstration tests (Demonstration II); (2) an evaluation of the two EDPs and tests for Pueblo; and (3) an evaluation of EDS packages and testing for Blue Grass. Technologies successful during Demonstration II tests will be candidates for use at Blue Grass. Thus, the third study could include evaluations of as many as four technologies. This report presents the committee's evaluation of the two EDPs and associated testing developed for Pueblo.

DESCRIPTION OF THE PUEBLO STOCKPILE

Agents

The principal unitary chemical agents in the U.S. stockpile are two nerve agents, GB and VX, and three related forms of blister agent, H, HD, and HT, which are also known as mustard. Only weapons containing HD and HT are stored at the Pueblo Chemical Depot in Colorado (see Table 1-2). The weapons are stored under ambient conditions at which the agent is primarily a liquid.

Mustard agent, which has a garlic-like odor, is hazardous on contact and as a vapor, is slightly soluble in water, and is very persistent in the environment. Table 1-3 lists some of the physical properties of mustard agents. As a result of aging, the actual composition of the agent stored at Pueblo may be somewhat different. In addition, the original composition varied slightly from one production lot to another. Tables 1-4 and 1-5 list the original typical compositions of HD and HT, respectively. Similar munitions at Johnston Atoll and at the Deseret Chemical Depot in Tooele, Utah, were found to contain a solid "heel," which did not flow from the shell.

---

[6]This technology was originally proposed to the ACWA program through a partnership between Parsons Infrastructure and Technology Group, Inc., and AlliedSignal, Inc. The latter has since been acquired by Honeywell, Inc.

[7]The Dialogue is a group of 35 stakeholder representatives, 4 of whom constitute the Citizens Advisory Technical Team (CATT). The Dialogue group is a mechanism for public involvement in decision making for the ACWA program; it was thoroughly described in the ACW I Committee's initial report (NRC, 1999). The CATT provides the Dialogue with more in-depth technical participation and interaction with the PMACWA staff, technology providers, and the ACW II Committee. CATT members often attend open sessions of the ACW II Committee, and the committee chair (or designates) regularly provides briefings on the progress of the committee's activities at meetings of the Dialogue.

TABLE 1-2 Munitions Containing HD and HT in the Pueblo Chemical Depot Stockpile

Munition Type	Model No.	Chemical Fill	Energetics	Configuration	Number
105-mm cartridge	M60	1.4 kg HD	Burster: 0.12 kg tetrytol Fuze: M51A5 Propellant: M67	Semifixed, complete projectile: includes fuze, burster. Propellant loaded in cartridge. Cartridges packed two per wooden box.	28,375
105-mm cartridge	M60	1.4 kg HD	0.12 kg tetrytol	Includes burster and nose plug but no fuze. On pallets.	355,043
155-mm projectile	M110	5.3 kg HD	0.19 kg tetrytol	Includes lifting plug and burster but no fuze. On pallets.	266,492
155-mm projectile	M104	5.3 kg HD	0.19 kg tetrytol	Includes lifting plug and burster but no fuze. On pallets.	33,062
4.2-inch mortar	M2A1	2.7 kg HD	0.064 kg tetryl Propellant: M6	Includes propellant and ignition cartridge.	76,722
4.2-inch mortar	M2	2.6 kg HT	0.064 kg tetryl Propellant: M6	Includes propellant and ignition cartridge.	20,384

Source: Adapted from U.S. Army, 1997.

TABLE 1-3 Physical Properties of Mustard Agents at Pueblo Chemical Depot

Agent Characteristic	HD	HT
Chemical formula	$C_4H_8Cl_2S$	60% $C_4H_8Cl_2S$, 40% T and impurities
Molecular weight	159.08	Not available
Boiling point (°C)	217	228
Freezing point (°C)	14.45	0 to 1.3
Vapor pressure (mm Hg)	0.072 at 20°C	
Volatility (mg/m^3)	75 at 0°C (32°F) (solid) 610 at 20°C (68°F) (liquid)	831 at 25°C (77°F)
Diffusion coefficient for vapor in air (cm^2/sec)	0.060 at 20°C (68°F)	0.05 at 25°C (77°F)
Surface tension (dynes/cm)	43.2 at 20°C (68°F)	44 at 25°C (77°F)
Viscosity (cS)	3.95 at 20°C (68°F)	6.05 at 20°C (68°F)
Liquid density (g/cm^3 at 20°C)	1.2685	1.22-1.24 (at ambient temperature)
Solubility (g/100 g of distilled water)	0.92 at 22°C (72°F); soluble in acetone, carbon tetrachloride, methyl chloride, tetrachloroethane, ethyl benzoate, ether	
Heat of vaporization (Btu/lb) (J/g)	190 82	Not available
Heat of combustion (Btu/lb) (J/g)	8,100 3,482	Not available

Sources: Adapted from NRC, 1993; U.S. Army, 1988.

TABLE 1-4 Original Nominal Composition of HD Mustard

Chemical Structure	Wt %
$ClCH_2CH_2SCH_2CH_2Cl$	89.2
$ClCH_2CH_2SCH_2CH_2SCH_2CH_2Cl$	4.7
$ClCH_2CH_2Cl$	2.4
$S(CH_2CH_2)_2S$	1.2
$S(CH_2CH_2)_2O$	0.5
$ClCH_2CH_2SCH_2CH_2CH_2Cl$	0.4
Unspecified	1.6

Source: Adapted from U.S. Army, 1997.

TABLE 1-5 Original Composition of HT Mustard

Chemical Structure	Wt %
$ClCH_2CH_2SCH_2CH_2Cl$	67.0
$(ClCH_2CH_2SCH_2CH_2)_2O$ [T]	22.2
$ClCH_2CH_2SCH_2CH_2OCH_2CH_2Cl$	4.5
$ClCH_2CH_2SCH_2CH_2SCH_2CH_2Cl$	3.0
$S(CH_2CH_2)_2S$	1.8
$S(CH_2CH_2)_2O$	0.5
$ClCH_2CH_2SCH_2CH_2OH$	0.4
$ClCH_2CH_2Cl$	0.4

Source: Adapted from U.S. Army, 1997.

Weapon Types

The mustard at Pueblo Chemical Depot is stored in artillery and mortar projectiles, which include a variety of other chemical compounds that must also be destroyed. The term "assembled chemical weapon" describes munitions that contain chemical agents. Mortars are typically stored with energetic components in place; projectiles may or may not contain bursters or fuzes. More detailed descriptions of these munitions are provided in Appendix A.

ROLE OF THE NATIONAL RESEARCH COUNCIL

The NRC has provided scientific and technical advice and counsel to the Army concerning the destruction of chemical weapons since the beginning of the CSDP. The history of this involvement was summarized in the first NRC report by the ACW I Committee (NRC, 1999) and will not be repeated here. The following discussion addresses only the role of the NRC in the ACWA program.

The PMACWA requested that the NRC conduct and publish an independent evaluation of the alternative technology packages representing a "total system solution" by September 1, 1999, a month before the Army's report to Congress was due. The NRC and DOD reached agreement on the Statement of Task in March 1997, and the study was officially begun on May 27, 1997. The committee decided to evaluate all seven technology packages that had passed the threshold requirements stipulated in the RFP, even though one was removed from further consideration by the Army during the course of the study. The statement of task did not require that the NRC recommend a best technology or compare any of the technologies with the baseline incineration process in use at two stockpile storage sites. Members of the ACW I Committee visited the demonstration sites prior to systemization of the unit operations in January 1999. However, in order to produce a report by September 1, 1999, data-gathering activities had to be terminated on March 15, 1999, prior to receiving the results of the demonstration tests. The committee's report, *Review and Evaluation of Alternative Technologies for Demilitarization of Assembled Chemical Weapons,* was submitted for peer review on May 1, 1999, and was released to the sponsor and the public on August 25, 1999 (NRC, 1999). This report found that the primary treatment processes could decompose the chemical agents with destruction efficiencies of 99.9999. However, major concerns for each technology package remained, including the adequacy of secondary treatment of agent hydrolysates and the primary and secondary treatment of energetic materials contained in the chemical weapons.

In September 1999, the PMACWA requested that the tenure of the committee be extended to review the results of the Demonstration I tests. The committee was asked to determine if and how the results affected the committee's commentary, findings, and recommendations, as well as the recommended steps required for implementation provided in the initial report. In October 1999, the committee began its evaluation of the results of the demonstrations and a determination of the impact of these results on its initial findings. The supplemental report was published in March 2000 (NRC, 2000). The tenure of the ACW I Committee was over at the end of March 2000, and a new committee, the ACW II Committee, was formed in May 2000 to evaluate (1) EDPs and testing for the Pueblo site, (2) EDPs and testing for the Blue Grass site, and (3) the Demonstration II tests. This report documents the ACW II Committee's review and evaluation of the EDPs for the Pueblo Chemical Depot.

STATEMENT OF TASK

The complete statement of task for the ACW II Committee study is given below:

> At the request of the DoD's Program Manager for Assembled Chemical Weapons Assessment (PMACWA), the NRC Committee on Review and Evaluation of Alternative Technologies for Demilitarization of Assembled Chemical Weapons will provide independent scientific and technical assessment of the Assembled Chemical Weapons Assessment (ACWA) program. This effort will be divided into three tasks. In each case, the NRC was asked to perform a technical assessment that did not include programmatic (cost and schedule) considerations.

Task 1

To accomplish the first task, the NRC will review and evaluate the results of demonstrations for three alternative technologies for destruction of assembled chemical weapons located at U.S. chemical weapons storage sites. The alternative technologies to undergo demonstration testing are: the AEA Technologies electrochemical oxidation technology, the Teledyne Commodore solvated electron technology, and the Foster Wheeler and Eco Logic transpiring wall supercritical water oxidation and gas phase chemical reduction technology. The demonstrations will be performed in the June through September 2000 timeframe. Based on receipt of the appropriate information, including: (a) the PMACWA-approved Demonstration Study Plans, (b) the demonstration test reports produced by the ACWA technology providers and the associated required responses of the providers to questions from the PMACWA, and (c) the PMACWA's demonstration testing results database, the committee will:

- perform an in-depth review of the data, analyses, and results of the unit operation demonstration tests contained in the above and update as necessary the 1999 NRC report *Review and Evaluation of Alternative Technologies for Demilitarization of Assembled Chemical Weapons* (the ACW report)
- determine if any of the AEA Technologies, Teledyne Commodore, and Foster Wheeler/Eco Logic technologies have reached a technology readiness level sufficient to proceed with implementation of a pilot-scale program
- produce a report for delivery to the PMACWA by July 2001 provided the demonstration test reports are made available by November 2000. (An NRC report delivered in March 2000 covered the initial three technologies selected for demonstration phase testing.)

Task 2

For the second task, the NRC will assess the ACWA Engineering Design Study (EDS) phase in which General Atomics and Parsons/Honeywell (formerly Parsons/Allied Signal) will conduct test programs to gather the information required for a final engineering design package representing a chemical demilitarization facility at the Pueblo, Colorado stockpile site. The testing will be completed by September 1, 2000. Based on receipt of the appropriate information, including: (a) the PMACWA-approved EDS Plans, (b) the EDS test reports produced by General Atomics and Parsons/Honeywell, (c) PMACWA's EDS testing database, and (d) the vendor-supplied engineering design packages, the committee will:

- perform an in-depth review of the data, analyses, and results of the EDS tests
- assess process component designs, integration issues, and overarching technical issues pertaining to the General Atomics and the Parsons/Honeywell engineering design packages for a chemical demilitarization facility design for disposing of mustard-only munitions
- produce a report for delivery to the PMACWA by March 2001 provided the engineering design packages are received by October 2000

Task 3

For the third task, the NRC will assess the ACWA EDS phase in which General Atomics will conduct test programs to gather the information required for a final engineering design package representing a chemical demilitarization facility at the Lexington/Blue Grass, Kentucky stockpile site. The testing will be completed by December 31, 2000. Based on receipt of the appropriate information, including: (a) the PMACWA-approved EDS Plans, (b) the EDS test reports produced by General Atomics, (c) PMACWA's EDS testing database, and (d) the vendor-supplied engineering design package, the committee will:

- perform an in-depth review of the data, analyses, and results of the EDS tests
- assess process component designs, integration issues, and overarching technical issues pertaining to the General Atomics engineering design package for a chemical demilitarization facility design for disposing of both nerve and mustard munitions
- produce a report for delivery to the PMACWA by September 2001 provided the engineering design package is received by January 2001.

SCOPE OF THIS REPORT

This report is the ACW II Committee's response to Task 2 of its Statement of Task (i.e., review and evaluate EDS documentation and testing developed for the destruction of chemical weapons at the Pueblo Chemical Depot). On the basis of the final schedule for the Pueblo EDS testing by the technology providers, the original delivery date of March 2001 was extended to July 15, 2001. This report will be produced in time to contribute to the Record of Decision (ROD) by the Office of the Secretary of Defense on a technology selection for the Pueblo site, which is scheduled for August 13, 2001, following satisfaction of NEPA requirements.

Because not all of the experimental test results in support of the EDPs were available as this report was being prepared, the committee was not able to review and evaluate them. However, committee members did attend status-review sessions organized by the PMACWA, and the committee was given access to all available draft reports by the technology providers.

ORGANIZATION OF THIS REPORT

This report has five chapters and four appendixes.[8] This first chapter has presented background information on the

---

[8]In this report, the findings and recommendations in the ACW I Committee's 1999 original report and its 2000 supplemental report are

ACWA program, the Pueblo Chemical Depot stockpile, and the NRC's involvement in the ACWA program. Chapter 2 discusses the results of hydrolysis studies on energetic materials sponsored by PMACWA in response to the ACW I Committee's original recommendations. Chapter 3 discusses the General Atomics EDP; Chapter 4 discusses the Parsons/Honeywell EDP. The results of testing completed in support of the EDPs and available to the ACW II Committee at the time this report was prepared are also reviewed. Chapter 5 summarizes the committee's evaluation of the two technology packages, presents some new general findings, and reevaluates the findings and recommendations of the ACW I Committee reports.

---

reviewed and updated. To avoid confusion, finding and recommendation designations have been adjusted as follows: no change to designations from the 1999 report; addition of (Demo I) to the designations from the 2000 supplemental report, e.g., Finding (Demo I) GA-1. Also, the findings and recommendations new to this report by the ACW II Committee include (Pueblo) in the designation.

2

Hydrolysis Tests of Energetic Materials

Both the General Atomics and the Parsons/Honeywell technology packages use caustic (base) hydrolysis as the initial step in destroying the energetic materials recovered from chemical weapons (Table 2-1). The technology packages use different secondary treatments of the products resulting from hydrolysis (i.e., the hydrolysates). General Atomics treats energetic materials hydrolysate by SCWO; Parsons/Honeywell uses biological treatment.

In the first ACW I Committee report, the committee expressed concern about the immaturity of the technology base for the hydrolysis of energetic materials. As the committee noted, although alkaline hydrolysis was effective in destroying energetic materials to 99.999 percent, the chemistry of the process was not well understood (NRC, 1999). The committee reiterated these concerns in the supplemental report (NRC, 2000). In response to these concerns, PMACWA initiated a multilaboratory test program during the EDS phase to clarify the chemical and engineering parameters for the efficient, safe alkaline hydrolysis of the energetic materials in assembled chemical weapons. In this chapter, the ACW II Committee briefly reviews disposal practices currently in use for energetic materials, briefly describes the hydrolysis treatment process for energetic materials, and describes PMACWA's program for engineering design testing of energetics hydrolysis. This is followed by an assessment of the status of the test program at the time this report was prepared, an evaluation of the results until that time, and a reassessment of the ACW I Committee's original findings.

CURRENT PRACTICES FOR THE DISPOSAL OF ENERGETIC MATERIALS

In the past, DOD disposed of a large percentage of unwanted munitions and the energetic materials they contained by an open burn/open detonation (OB/OD) process, an environmentally undesirable approach that has already been banned in Europe. DOD has begun to work toward minimizing OB/OD as a means of disposal (JOCG, 2000).

When an item is identified as surplus, it is transported to a site at which it will either be destroyed immediately or stored until it can be destroyed. Most of the sites that receive munitions for destruction are Army depots such as Hawthorne Army Depot (Nevada), Sierra Army Depot (Nevada), and Hill Air Force Base (Utah) (Mitchell, 1998). Other sites that sometimes receive munitions for experimental purposes include the Naval Surface Warfare Center (NSWC) Indian Head Division (Maryland) and the Nevada Test Site of the Department of Energy.

Frequently discussed alternatives to OB/OD include processes known as resource reclamation and recycling (R3). These R3 processes are designed to either reclaim and recycle valuable metals from obsolete surplus ordnance or

TABLE 2-1 Nominal Composition of Energetic Materials Used in Chemical Munitions

Energetic Material	Composition
Tetryl	2,4,6 trinitrophenylmethylnitramine
Tetrytol	70% tetryl / 30% TNT
Composition B	60% RDX / 39% TNT / 1% wax
Composition B4	60% RDX / 39.5% TNT / 0.5% calcium silicate
M28 propellant	60.0% nitrocellulose / 23.8% nitroglycerin / 9.9% triacetin / 2.6% dimethylphthalate / 2.0% lead stearate / 1.7% 2-nitrodiphenylamine
M8 propellant	52.15% nitrocellulose / 43% nitroglycerin / 3% diethylphthalate / 1.25% potassium nitrate / 0.6% ethyl centralite
M1 propellant	84% nitrocellulose / 9% dinitrotoluene / 5% dibutyl phthalate / 1% diphenylamine / 1% lead carbonate

Source: Bonnett, 2000.

reclaim potentially valuable energetic components such as TNT, RDX, and HMX for reuse (Garrison, 1994). However, many of the R3 processes are still being evaluated for economic feasibility in a number of DOD demonstration programs (Newman et al., 1997; Marinkas et al., 1998; Goldstein, 1999). According to Mitchell (1998), "In 1998, approximately 60% of the 100,000 tons of demilitarization surplus ordnance were disposed in a way which enabled at least some of the material to be recovered and recycled."

Some energetic materials do not lend themselves to recovery and recycling, either because the economics of the process are unfavorable or because the material properties are unfavorable. Nitrocellulose-based propellants and materials containing nitrate ester plasticizers are not suitable feedstocks for an R3 program because of their long-term instability. Compositions containing these ingredients always include a stabilizer to prevent catastrophic self-heating as the materials age. However, the degradation of the propellants and the presence of impurities in aging energetics of this type make them poor candidates for the economical recovery of energetic components. These materials are still destroyed by OB/OD. Research has been done to evaluate the potential of demilitarized gun propellants for a variety of uses, such as sensitizers for commercial slurry explosives and boiler fuels (Machacek, 2000).

The demilitarization of small items, such as igniters and fuzes, is routinely accomplished in an APE-1236 furnace, a rotary kiln in which the devices are heated until the energetic material decomposes thermally. The amount of material that could be recovered from these items is small, and the energetic materials themselves, especially detonators, are often quite sensitive. Because of their sensitivity, attempting to disassemble the items would be more hazardous than disassembling main-charge explosives. Therefore, these items are either intentionally "functioned" (i.e., actuated) or thermally decomposed.

Alternative technologies to OB/OD for items that contain energetic materials not worth recovering are being explored but are not widely used. Confined burning, a process in which the gaseous and condensed products of combustion can be captured and treated before release, is being used at some sites around the country. Hydrolysis of energetics as a means of disposal is being used at the Hawthorne Army Depot.

Several other technologies (e.g., molten-salt destruction) are being used at research and development sites (e.g., Eglin Air Force Base and Strauss Avenue Thermal Treatment Plant) to destroy energetic materials, but these technologies are not an integral part of DOD's plan for the demilitarization of obsolete munitions.

CAUSTIC HYDROLYSIS OF ENERGETIC MATERIALS

Caustic hydrolysis of energetic materials has been investigated as an alternative technology to the OB/OD method.

Newman (1999)[1] published a review of the known chemistry of caustic hydrolysis of energetic materials used in assembled chemical weapons, and recent work on the destruction of aromatic nitro compounds (TNT and tetryl) by alkaline hydrolysis has been reported (Bishop et al., 2000).

The chemistry of caustic hydrolysis takes advantage of the susceptibility of the functional groups commonly found in energetic materials to attack by hydroxide ion, which yields products that are essentially nonenergetic. Caustic hydrolysis decomposes energetic materials to organic and inorganic salts, soluble organic compounds, and various gaseous effluents. Partial hydrolysis of some energetic materials, particularly materials with aromatic ring systems, may lead to ill-defined oligomeric materials with low solubility in either aqueous or organic solvents.

The rate of reaction depends on, among other things, the concentration of the energetic compound in solution or, for heterogeneous reactions, on the surface area of the solids being hydrolyzed. An important factor in determining the rate of destruction is the phase of the compound in the hydrolysis reactor. The compounds of interest may be divided into three classes:

- compounds that are liquids at normal reactor temperatures (e.g., 2,4,6-TNT and nitroglycerin)
- compounds that are solids at normal reactor temperatures (e.g., RDX and tetryl)
- polymeric materials (usually nitrocellulose)

TNT has low solubility in aqueous solutions and forms an emulsion with hot caustic solution. Thus, because the TNT is molten, the size of the droplets in the emulsion is determined not by the size of the granules in the original feedstock but by the degree of agitation in the hydrolysis reactor, as well as the presence of any surfactant. For Composition B, the size of the RDX particles in the reactor will reflect the size that was used in manufacturing the Composition B.[2] During manufacturing, a small fraction of the RDX dissolves in molten TNT, but the remainder is suspended in the TNT matrix. Therefore, when the TNT is remelted, the original RDX particles can be recovered. Thus, the RDX particle size does not depend on the size of the Composition B pieces fed into the reactor but on the size of the original RDX particles mixed into the TNT, typically between 10 μm and 1 mm. The particle distribution may be skewed toward the larger particles because the smaller particles dissolve more rapidly in the TNT.

---

[1]This information can also be found in condensed form in Appendix E of the initial ACW I Committee report (NRC, 1999).

[2]Composition B contains 1 percent wax. Depending on the nature of the wax, some long-chain fatty acids may be present, which act as surfactants. Hydrolysis of the plasticizers in M28 propellant may also release phthalate salts, which can aid in the emulsification of TNT when M28 and Composition B are processed together.

Tetryl, like RDX, is a solid at the temperatures in the hydrolysis reactor. For the neat tetryl in burster charges, the grain size depends on the extent to which the pressed explosive charges have been processed prior to being added to the hydrolysis reactor. The case of tetryl in tetrytol is quite different. TNT and tetryl are very similar chemically, so the solubility of tetryl in molten TNT is quite high (82 g/100 g TNT at 80°C [176°F]; 149 g/100 g TNT at 100°C [212°F]) (Kaye and Herman, 1980). Tetryl in tetrytol is mostly dissolved in the TNT phase, so the rate of dissolution and subsequent reaction in the hydrolysis medium depends mainly on the TNT/tetryl droplet size and not on the particle size of the tetryl that was originally used to make the tetrytol.

Polymeric nitrocellulose is solubilized by the degradation of the glycosidic linkages along the polymer chain. Nitrocellulose-based propellant grains contain other components that are released as the nitrocellulose breaks down. Even before it is completely broken down, infiltration of caustic into the partially decomposed propellant grains may allow the nitroglycerin inside the grains to come in contact with the caustic medium and react.

In an effort to increase the reaction rate of the energetic materials in the hydrolysis reactor, strong caustic solutions (pH greater than 12), elevated temperatures (60°C to 155°C [140°F to 311°F]), and elevated pressures (up to 14 atm) have been used. The hydrolysis reaction is exothermic, so process controls are necessary to maintain the reactor temperature and respond to thermal excursions in order to prevent a runaway reaction. Because upsets are always possible, the reactors and containment buildings must be designed to contain the maximum credible explosive event.[3]

A thorough understanding of fundamental requirements for the hydrolysis of energetic materials is essential to the design and operation of a chemical agent demilitarization facility, where high levels of engineering controls are necessary to ensure the safe disposal of chemical agent and the prevention of process upsets throughout the facility. The ACW I Committee report indicated that four issues surrounding the caustic hydrolysis of energetic materials must be addressed (NRC, 1999):

- determination of the steps required for removing the energetic material from the munitions and reducing it to the appropriate particle size
- determination of safe operating parameters for heterogeneous mixtures of energetics, metals, and contaminants

- development of process controls and equipment to minimize the accumulation of precipitates and minimize the effects of an accident
- determination of the scale-up parameters to meet the destruction requirements for diverse munitions

OVERVIEW OF THE TEST PROGRAM

In response to the challenges listed above, the PMACWA devised an EDS test plan that addresses some of these issues. The start of testing was delayed from August 2000 to December 2000, and the tests were not to be completed until the end of March 2001. The responsibility for coordinating the program was assigned to U.S. Army Tank-Automotive and Armaments Command (TACOM) Armament Research, Development and Engineering Center (ARDEC) at Picatinny Arsenal, New Jersey.

The *Picatinny Test Plan Requirements* describes objectives, planned testing, and team member responsibilities for the EDS test program (Bonnett, 2000). The main objectives of the test plan and some of the organizations involved are summarized below:

- Picatinny Arsenal and Holston Army Ammunition Plant (HAAP) are responsible for determining and defining optimum operating parameters for the hydrolysis of all energetic materials contained in munitions at the Pueblo Chemical Depot and the Blue Grass Army Depot.
- Los Alamos National Laboratory (LANL) and NSWC Indian Head Division (Yorktown, Virginia) are responsible for performing bench-scale tests to address ACW I Committee concerns about the solubility of energetics in alkaline solutions, the simultaneous processing of different types of energetics, and the reduction to the proper particle sizes for operation.
- TACOM ARDEC is responsible for incorporating data generated from bench-scale tests into full-scale production processes at HAAP to demonstrate the hydrolysis operations.
- The Pantex Plant is responsible for production of tetrytol hydrolysate.
- The Radford Army Ammunition Plant (RAAP) is responsible for production of M28 simulant hydrolysate.

Because the EDS test program had not been completed at the time this report was submitted, only interim results were evaluated.

Testing at the Holston Army Ammunition Plant

The reactor at HAAP has the following characteristics:

- a 2,000-gallon, glass-lined, jacketed reactor vessel
- a recirculation loop, unheated and uninsulated, that

---

[3] A determination of the maximum credible explosive event is made by considering the probabilities of accidental violent reactions (e.g., rapid burning or detonation of energetic materials) and the resulting damage and hazards associated with each event. The resulting hazards define the appropriate hazard mitigation and control strategies that are used to minimize the impact of an accidental explosion in a facility in order to protect people and property.

reenters the reactor vessel at the bottom
- a dual-flight, Hastelloy C, variable-speed agitator in the reactor vessel
- a condenser/scrubber for off-gases
- pH-, temperature-, and flowmeters

The energetic feeder is a single-screw, loss-in-weight feeder with a 500 lb/hr capacity. The control system uses programmable logic control (PLC) with remote and local control capabilities. Energetic material will be fed into the reactor as a dry solid and screened as it is added to the feeder hopper to prevent particles larger than 0.5 inches in diameter from entering the reactor.

Demonstration runs are being conducted at various operating conditions for seven different energetic materials and mixtures (Bonnett, 2000):

- M1 propellant
- M8 propellant
- tetrytol
- M28 propellant
- Composition B
- Composition B4
- M28 propellant/Composition B4 (86/14 weight-percent ratio based on amounts in an M55 rocket)

The primary process parameters to be studied are (1) reactor residence time, (2) energetic feed rate, (3) reaction temperature, (4) caustic concentration, and (5) agitation speed.

Although the munitions at Pueblo Chemical Depot contain only M1 propellant, M8 propellant, and tetrytol, the EDS tests with M28 propellant and Composition B high explosive are included in this chapter for completeness and because the tests are closely related. Composition B has been studied in anticipation of demilitarization operations at the Blue Grass, Kentucky, storage site; M28 propellant was used for testing the ERH, a component of the General Atomics technology package (see Chapter 3).

An acceptable energetic feed rate will be based on (1) control of the heat generated to avoid thermal runaway, (2) completeness of the reaction, (3) achievable and effective settings for agitation speed, (4) elimination/minimization of foam formation, and (5) elimination/minimization of the production of undesirable by-products (Bonnett, 2000). The Army expects the time required for addition of the energetic materials to be inconsequential when compared with the total processing time required for the hydrolysis reaction. Therefore, a broad range of feed rates will be consistent with operational safety.

Chemical sensors will be used and control strategies developed to ensure that all energetic materials are fully hydrolyzed and that process and operator safety is maintained. Other objectives of the system performance evaluation are summarized below (Bonnett, 2000):

- to demonstrate performance of the control sensors and the logic programming for the sensors
- to determine the efficacy of process cooling and temperature control
- to determine the efficiency of agitation by the impeller and the recirculation loop
- to determine the efficacy of steps taken to control foam formation
- to determine condenser performance
- to determine maintenance requirements
- to develop a contingency for sudden shutdown

Hydrolysate samples collected during tests will be used in the following ways:

- to characterize fully hydrolysate products as a function of time and reactor operating parameters (e.g., temperature, residence time, and caustic concentration)
- to determine if any energetic materials are generated as by-products (e.g., picric acid)
- to evaluate the stability of hydrolysate for post-treatment processing
- to determine final product characteristics that influence post-treatment technology (e.g., pH, solid content, particle size, and homogenization)

At the conclusion of each test run, the interior surfaces of the reactor will be visually inspected and samples of residue, if present, will be collected and characterized. A detailed preventive maintenance program will be developed to minimize the possibility of incidents during the cleanup of accumulated precipitates. Materials of construction will be investigated for alkaline and acid resistance. The rate of buildup of potentially energetic by-product salts will be assessed. The type and frequency of maintenance will be determined.

Bench-scale Tests at Los Alamos National Laboratory

Researchers at LANL are characterizing the hydrolysis reactions at bench scale by analyzing the hydrolysate and gaseous effluents generated by the processing of energetic materials. The concentration of caustic will be varied (12, 20, 25, and 30 weight percent) in these tests to determine its influence on destruction efficiency and residence time. LANL will also investigate the feasibility of hydrolyzing mixtures of energetics, the effects of particle size on reaction rates, the formation and growth of crystals in the hydrolysate, and the feasibility of mixing various energetic hydrolysates (Bonnett, 2000). These data will be used in developing requirements for size-reduction equipment and the methodologies for handling incoming energetic materials during full-scale processing.

LANL and NSWC are also investigating the feasibility of

processing mixtures of energetic materials found in a single type of munition:

- M1 propellant/tetrytol (105-mm M2 cartridge)
- M8 propellant/tetryl (4.2-inch, M2 cartridge)
- M28 propellant/Composition B4 (115-mm, M55 rocket)

The actual ratios of propellant to burster explosive in the munitions will be used. Hydrolysate products are also being analyzed for picrate, as recommended in the ACW I Committee report (NRC, 1999). LANL has hydrolyzed the following combinations with no major perturbations: tetrytol (TNT and tetryl); cyclotol (RDX and TNT); octol (HMX and TNT); and nitrocellulose, nitroglycerin, nitroguanidine, triple-base propellant, and HMX (Bishop, 2000). Processing perturbations such as foaming were managed and controlled using well-known engineering techniques.

Bench-scale Tests at the Pantex Plant

Hydrolysis experiments at Pantex have shown that cyclotol (70 percent RDX and 30 percent TNT) and tetrytol (70 percent tetryl and 30 percent TNT) reacted within 1 hour and 3 hours, respectively, in 6 to 12 percent caustic. The metric for the completion of the reaction is the disappearance of solid material in the reactor (Belcher, 2000). The reaction time for the tetrytol was probably less than 3 hours, because the functional groups in the tetryl molecule, which are similar to those in TNT and RDX, should react at similar rates. However, only a lower bound on the rate of tetrytol destruction is available, because no observations were made before 3 hours had elapsed. Observations made after 1 hour for cyclotol showed that all solids had been consumed.

Bench-scale Tests at the Naval Surface Warfare Center

The NSWC will conduct calorimetric studies to determine the heat of reaction for hydrolysis reactions with various concentrations of caustic. This information will be used to develop strategies for reaction controls and to prevent runaways and upsets (Bonnett, 2000).

The following reaction parameters are being determined for each energetic material by accelerating rate calorimetry:

- temperature of the maximum self-heating rate
- dependence of reaction rate on pressure and temperature
- rate of pressure and temperature increase
- heat of reaction
- moles of gas evolved per unit mass of energetic material
- activation energy of the reaction
- reactor cooling requirements

This information will be useful for numerical modeling and simulation of the hydrolysis reaction process.

Hydrolysate Production at the Radford Army Ammunition Plant

Prior to Demonstration I, some attempts had been made to hydrolyze energetic materials on a large scale. RAAP (along with the Pantex Plant) produced the hydrolysates used during Demonstration I and the EDS tests in the ACWA program. RAAP was also expected to produce hydrolysate from M28 surrogate for the EDS program.

RAAP had manufactured M28 surrogate propellant specifically for the preparation of hydrolysate. For environmental reasons, the surrogate did not contain lead stearate, which is normally included in M28 propellant as a burn-rate modifier. The propellant was prepared in grains in the shape of right circular cylinders, 1/16 inch in diameter by 1/16 inch long.

Some of the problems that might be encountered in a large-scale operation were illustrated by a recent upset at RAAP. On October 14, 2000, hydrolysate from M28 surrogate propellant was being prepared when the piping of the recirculation loop ruptured, causing significant damage to the equipment (described later in this chapter).

PROGRAM STATUS

Results of Tests at the Holston Army Ammunition Plant

Because the start of the EDS test program on energetics hydrolysis was delayed, the testing at HAAP had generated only limited data at the time this report was prepared. Energetic materials representative of the materials in the Pueblo stockpile had not yet been tested in the full-scale reactor at HAAP. As of January 1, 2001, only two Composition B hydrolysis runs had been completed. In one run, 200 lb of Composition B were hydrolyzed, and in the other, 500 lb were hydrolyzed. A detailed analysis of the hydrolysate composition as a function of time during these runs had not been completed by February 1, 2001; however, useful information was generated about the systemization (preoperational testing) of a full-scale hydrolysis reactor. At this point, the committee can comment only on the test plan and the preliminary results from these two runs. The committee believes that the test plan is well designed to determine acceptable parameters on the full-scale reactor at HAAP.

Systemization of the 2,000-gallon hydrolysis reactor at HAAP was completed within 4 months. Composition B, which is produced at HAAP and is readily available, was chosen for the initial experiments. The disposal of hydrolysate is covered under existing permits for handling waste from the production of Composition B. Because foaming is difficult to control in Composition B, the two test runs with this material provided a good test of the efficacy of measures designed to control foaming.

The problems that occurred were typical of any start-up operation. For example, the Acrison feeder, which is used to deliver Composition B to the reactor, failed to operate because of a software error when it was loaded with 500 lb of material. Water inadvertently added to the reactor overflowed into the dump tank and the secondary containment. The mass flowmeter in the recirculation loop malfunctioned. Despite these problems, feed rates of more than 490 lb/hr were maintained during the addition phase of the run, and the exothermicity was controllable. No components of Composition B (RDX and TNT) were detected at the end of the run in samples taken from the recirculation loop. Although complete characterization is still required, preliminary results to date have indicated that caustic hydrolysis is an acceptable technology for destroying energetic materials in a full-scale operation.

The energetics hydrolysis reactor at HAAP is larger than the reactors in either EDP for Pueblo. The proposed systems will be suitable for the small quantity of energetic materials to be treated at Pueblo, but conclusions based on the HAAP experience (a 2,000-gallon reactor) must be reconsidered to account for the intended feed rate and energetic materials loading in pounds/gallon/hour for the 200-gallon reactor at Pueblo.

Results of Tests at Los Alamos National Laboratory

Characterizing the progress of the reaction is a prerequisite for designing a process for destruction of energetic materials by means of hydrolysis. The progress of the reaction can be followed by observing the disappearance of the initial feed. The studies at LANL are designed to assess the risk of untreated energetic materials in solution after hydrolysis has been completed. The solubilities in water of RDX and TNT at 60°C (140°F) have been reported in the literature to be approximately 300 ppm and 600 ppm, respectively (Gibbs and Popolato, 1980). LANL has determined the solubility of HMX at 90°C (194°F) to be approximately 150 ppm (Bishop, 2000). Because any dissolved energetic material is rapidly hydrolyzed (RDX half-life <1 s, HMX half-life of 0.92 min at 25°C (72°F) and 1.5 M NaOH), the absence of any solids is a valid indicator that all of the original energetic material has dissolved and reacted. The solubilities of the hydrolysate products and their cumulative effects are being investigated at LANL. These findings will guide the pilot-plant and full-scale demonstration runs.

Test results at LANL indicate that reliable mass balances for the hydrolysis reactions of aromatic nitro compounds, such as TNT and tetryl, are difficult to obtain. A significant fraction of the carbon in the hydrolysis of TNT and tetryl ends up as a mixture of high molecular weight compounds (mol. wt. = 3,000 to 30,000) that do not appear to be energetic (as evaluated by a simple hammer test) but may present other problems in subsequent processing steps. For example, because they are viscous, they might clog pumps or pipes. Also, they might separate from the aqueous phase of the hydrolysate and affect the reaction time for complete hydrolysis.

The researchers at LANL recognize the concerns of the ACW I Committee about the formation of picric acid in the presence of lead, particularly the lead stearate in the M28 propellant, which could lead to the formation of lead picrate, a very sensitive explosive (Bishop, 2000). LANL characterized TNT and tetrytol hydrolysate for picric acid by high-pressure liquid chromatography and differential scanning calorimetry and found no evidence of picric acid or any other known energetic material (Bishop, 2001). Although the primary concern was the lead stearate in M28 propellant, General Atomics proposes hydrolyzing other lead-containing components, such as lead azide in fuzes, simultaneously with M28 propellant in the ERH. (In the Parsons/Honeywell design, fuzes would not be treated in the hydrolysis reactors but would be heated in an energetics rotary deactivator [ERD] until they deflagrate or detonate [see Chapter 4].) The LANL researchers believe that the lead stearate will likely form products such as lead hydroxide, which is more insoluble than lead picrate (Bishop, 2001). Based on the studies at LANL, the committee believes secondary treatment of the hydrolysate will have to treat any insoluble products that may form during hydrolysis or subsequent processing of the waste streams from the demilitarization plant. The toxicity of products formed during the entire processing cycle of the waste stream must be considered.

The EDS hydrolysis testing at LANL is still under way. The researchers will continue to investigate the hydrolysis of all of the energetics for the neutralization of the assembled chemical weapons stockpile. They also plan an online analysis of the gas-phase products of hydrolysis, an analysis of the solid residue (including testing for the presence of energetic materials), and an analysis of the gases evolved from the hydrolysate during storage. Further studies of pH neutralization and an analysis of lead products may also be undertaken. The results will add to the technology base for the hydrolysis of energetic materials.

Results of Tests at the Naval Surface Warfare Center

The experimental database generated thus far includes completed accelerating rate calorimetry runs of neat energetic materials and runs of a mixture of tetrytol in a 30 weight percent solution of NaOH (Bishop, 2000). The results of these experiments are intended to provide a better understanding of the controls necessary to prevent runaway reactions. It would be premature for the committee to draw conclusions based on the limited results available so far.

Analysis of an Incident at the Radford Army Ammunition Plant

The M28 surrogate was hydrolyzed in an existing tank that had previously been used only for the preparation and storage of aqueous NaOH solution. Although the tank had an agitation mechanism, it was suspected to be inadequate for complete and efficient hydrolysis. Additional agitation was provided by an external recirculation loop that accepted material from the bottom of the tank and returned it to the top. The external recirculation loop had steam trace heating and two parallel pathways fitted with closed impeller pumps that were not designed to handle slurries.

In preparation for the hydrolysis, the contents of the tank were brought to the proper NaOH concentration (12 weight percent) by diluting the appropriate amount of 50 percent caustic. The temperature was adjusted to 91°C (195°F) using internal steam (40 psig) coils. This temperature was maintained overnight. On October 13, particles of M28 were added incrementally over several hours through a hopper that directed them onto a slanted screen, where they were supposed to remain until they were digested. However, the agitation suspended the propellant grains in the caustic solution, and the particles later settled on the bottom of the tank at the intake for the recirculation loop.

At 8:15 A.M. on October 14, 2000, one of the two pumps in the recirculation loop was turned on, but it stalled and had to be shut down. This pump had frequently stalled even when only caustic NaOH solution was present, so nothing seemed unusual. At 8:25 A.M., the other pump was started; at 9:50 A.M., it was found to have been pushed apart from overpressurization resulting from a blockage. Both valves on the recirculation line, one near the bottom of the tank and one at the pump suction, were then closed off, effectively stopping the recirculation. The steam trace line heating was continued. At 11:05 and 11:30 A.M., the recirculation loop piping was overpressurized, thereby causing the pipe to split and the flanges to separate. Partially digested propellant spilled onto the floor, but the propellant did not ignite. The system was cooled; contents of the tank were left in place, pending a decision on how to remove them safely.

The chain of events leading up to the pressure rupture of the pipes appears to have started with the ingestion of partially decomposed propellant grains into the recirculation line. The recirculation pumps became clogged with the grains, and when the valves were closed at the intake and pump ends of the lower leg of the recirculation loop, the mixture of caustic and propellant was trapped in the piping and there was no way for the gases generated by the decomposition of the propellant to escape. The reaction was sustained by the combination of heat generated by the decomposition and heat supplied by the steam lines. The buildup of pressure eventually led to separation of the flanges and rupture of the piping. This is an important illustration of the potential hazards of reactions with energetic materials. These hazards must be controlled and accident scenarios must be considered.

Some of the lessons learned from the RAAP that are germane to the HAAP experiments include pump design, recirculation intake location, screen mesh size for controlling migration of undesired energetics, and heat transfer. Pumps need to have an open design capable of handling slurries. Any recirculation loop needs to be located away from the bottom of the reactor vessel (or any other dead zone) so that undesired chunks and precipitates are not entrained through small-flow geometries. Screen mesh size and screen location need to be chosen carefully to ensure that chunks are controlled and the screen is kept open enough to maintain flow. Heat transfer data (i.e., inputs and outputs) need to be integrated into a control system that links process perturbations with variables subject to manipulation. Sensors should be identified that can confirm the following:

- the flow in narrow geometries (such as a nonintrusive flowmeter on the recirculation loop)
- the rate of temperature rise in locations that correspond to anticipated hot spot formation (such as thermocouples located at or close to the minimal clearance points between the wall and agitator)
- the agitator speed in the reactor (i.e., a tachometer)

Process control setpoints and limits for acceptable operation should then be established, along with control algorithms that implement corrective action if process upsets or perturbations are detected.

The severity of the incident might have been mitigated if consideration had been given to the reaction that was taking place between the propellant and the caustic. Failure to stop the steam trace heating on the recirculation loop helped to sustain the temperature needed for the reaction to continue, and closing the valves at both ends of the segment of the loop below the tank ensured that the gases produced would build up pressure.

Although the incident at RAAP is unlikely to occur at Holston because the intake position and the type of pumps used are different, blockage could be caused by something else. For example, one of the compounds to be hydrolyzed at Holston is TNT, which has a melting point of 82°C (180°F). The intended reactor temperature for the hydrolysis experiments is between 85°C (185°F) and 95°C (203°F). Because the recirculation loop is neither insulated nor heated in any way, TNT might cool and crystallize in the piping when the reactor is being run at the lower temperature.

None of the munitions at Pueblo contains M28 propellant; therefore, this incident has no direct bearing on planned disposal activities at Pueblo. However, the incident does show that even though the maximum credible event may not result from every process upset, sound engineering practices must always be used. This will reduce the likelihood of an accident and mitigate the consequences of accidents that do occur. The incident also highlights the need for training personnel involved in such operations to become aware of all possible hazards.

SUMMARY ASSESSMENT

The EDS energetics hydrolysis test program was instituted to address the ACW I Committee's findings and recommendations on the hydrolysis of energetic materials. The test program was delayed and had not progressed very far by the time this report was submitted for review. Consequently, only limited information was available. The plan for the program, however, addresses all of the findings and recommendations of the ACW I Committee (NRC, 1999, 2000) with the exception of the hydrolysis of energetics contaminated by agent.

The data from pilot-scale hydrolysis of contaminated energetics confirm that common energetic materials (e.g., nitro compounds, nitramines, and nitrate esters) can be effectively and safely converted to nonenergetic products by the action of caustic at elevated temperature (Bonnett, 2001). However, the colocation of operations involving chemical agents and explosives increases the safety concerns raised by any operation involving energetics. Under normal circumstances, process upsets and deviations from standard operating procedures occasionally lead to unplanned explosions or detonations. For this reason, stringent regulations have been promulgated to isolate other operations from explosives-handling operations. The value of these safeguards was apparent in the incident at RAAP. Similarly, consideration should be given to the proximity of operations involving agent and operations involving energetics in facility designs. Even as benign an event as a leak or spill of caustic solution charged with energetics could seriously disrupt the ingress and egress of personnel at nearby facilities that have nothing to do with the energetics operation until the spill has been cleaned up and an incident investigation completed. Process upsets and unplanned events that can be tolerated in a nonchemical-weapons environment are not acceptable in a chemical-weapons destruction facility.

In a facility for demilitarization of assembled chemical weapons, otherwise minor upsets cannot be tolerated. Therefore, a thorough understanding of all aspects of the energetics hydrolysis process will be essential. Experience thus far with tests at HAAP suggests that the concerns involving the immaturity of the hydrolysis process for energetics, which were identified in the findings and recommendations of both ACW I Committee reports, have not been fully addressed. The ACW II Committee's concern is focused on the possible impact of process upsets on the facility, rather than on the adequacy of the hydrolysis process to neutralize energetic materials.

The completion of the EDS test program is expected to provide a much more complete understanding of the hydrolysis process, control systems, maintenance requirements, and other considerations necessary for determining the applicability of this technology to assembled chemical weapon demilitarization. However, EDS experiments being conducted with tetrytol, which contains TNT, will not be representative of the hydrolysis of neat tetryl. The TNT, which is molten at the reaction temperature, will change the physical state of the tetryl when tetrytol is the feedstock. The relationship between the rate of destruction of tetryl and the granulation of the feedstock material from neat tetryl boosters has not yet been established.

The ACW II Committee will continue to monitor the evolving test data and results from the various locations conducting ACWA EDS tests on energetics hydrolysis, including such issues as the characterization and treatment of off-gases. Further evaluation of this developing information will be made by the committee in a forthcoming report (to be published in 2002) on the EDS program for proposed alternative technologies for the Blue Grass site.

Previous Findings and Recommendations of the ACW I Committee

In this section, the findings and recommendations on energetic hydrolysis from the two ACW I Committee reports are reviewed to determine the extent to which they are still valid as a result of EDS testing (NRC, 1999, 2000).

Review of Findings and Recommendations from the 1999 Initial ACW I Committee Report

Finding GA-2. Hydrolysis of energetics at the scales proposed by the technology provider is a relatively new operation. Chemically, it is possible to hydrolyze all of the energetic materials; however, the rate of hydrolysis is limited by the surface area and, therefore, depends on particle size. (Smaller particles are more desirable because they have a higher surface-to-volume ratio.) The proposed method of removing and hydrolyzing the energetics appears to be reasonable, but further testing is required to determine the hydrolysis rates and to confirm that throughput rates can are achieved.

This finding is being addressed by the ACWA EDS program.

General Finding 2. The technology base for the hydrolysis of energetic materials is not as mature as it is for chemical agents. Chemical methods of destroying energetics have only been considered recently. Therefore, there has been relatively little experience with the alkaline decomposition of ACWA-specific energetic materials (compared to experience with chemical agents). The following significant issues should be resolved to reduce uncertainties about the effectiveness and safety of using hydrolysis operations for destroying energetic materials:

- the particle size reduction of energetics that must be achieved for proper operation
- the solubility of energetics in specific alkaline solutions
- process design of the unit operation and the identification of processing parameters (such as the degree of

agitation and reactor residence time) necessary for complete hydrolysis
- the characterization of actual products and by-products of hydrolysis as a function of the extent of reaction
- the selection of chemical sensors and process control strategies to ensure that the unit operation following hydrolysis can accept the products of hydrolysis
- development of a preventative maintenance program that minimizes the possibility of incidents during the cleanup of accumulated precipitates.

This finding is being addressed by the ACWA EDS program. The assessment of the particle size reduction needed for proper operation was limited to a single particle size, which was acceptable for the operation. No attempt to identify an optimum particle size is included in the program.

General Finding 3. The conditions under which aromatic nitro compounds, such as trinitrotoluene (TNT) or picric acid, will emulsify in the aqueous phase and not be completely hydrolyzed are not well understood. Therefore, this type of material could be present in the output stream from an energetic hydrolysis step.

Complete destruction has now been demonstrated.

General Finding 4. The products of hydrolysis of some energetic materials have not been characterized well enough to support simultaneous hydrolysis of different kinds of energetic materials in the same batch reactor.

This finding is being addressed by the ACWA EDS program.

General Recommendation 5. Whatever unit operation immediately follows the hydrolysis of energetic materials should be designed to accept emulsified aromatic nitro compounds, such as TNT or picric acid, as contaminants in the aqueous feed stream (see General Finding 3).

This recommendation is still valid until the EDS testing program is completed and the results indicate otherwise.

General Recommendation 6. Simultaneous processing of different types of energetic materials should not be performed until there is substantial evidence that the intermediates formed from the hydrolysis of aromatic nitro compounds will not combine with M28 propellant additives or ordnance fuze components to form extremely sensitive explosives, such as lead picrate (see General Finding 4).

This recommendation has been addressed to the extent that the LANL study shows the absence of picrate in hydrolysis products. As yet, mixtures of energetics have not been effectively addressed in the EDS test program, although such tests are planned. Until those tests are completed, mixtures of energetic materials should not be hydrolyzed on a large scale.

Review of Findings from the ACW I Committee Supplemental Report (NRC, 2000)

Finding (Demo I) GA-1. Testing on the hydrolysis of energetic materials contaminated with agent will be necessary before a full-scale system is built and operated.

This finding is not being addressed by the ACWA EDS program. The committee notes that integration concerns such as this should be addressed as soon as practicable to minimize delays during systemization of the disposal facility (see General Finding [Pueblo] 3 in Chapter 5).

New Findings and Recommendations

Finding (Pueblo) EH-1. Alkaline hydrolysis can be an effective and safe method for destroying energetic materials at Pueblo Chemical Depot. There appear to be no insurmountable obstacles to using this technology to destroy the energetics in assembled chemical weapons.

Finding (Pueblo) EH-2. Results from the energetics hydrolysis test program thus far have shown that hydrolysis rates are consistent with the proposed designs for overall throughput rates necessary to meet the current disposal schedule for the Pueblo stockpile.

Finding (Pueblo) EH-3. The hydrolysis of neat tetryl from burster charges is not being tested. Tests with tetrytol, which contains TNT, will not be representative of the hydrolysis of neat tetryl.

Finding (Pueblo) EH-4. Although the EDS energetics hydrolysis test program addresses many of the issues related to effective destruction of energetic materials from assembled chemical weapons, the tests are based on a predetermined granulation of the feedstock and will not provide information for determining the optimum granule size for disposal operations at Pueblo. The tests will provide information for only one granulation size and will not show the relationship between destruction rate or efficiency and particle size.

Finding (Pueblo) EH-5. Mass balances for most of the data from bench-scale hydrolysis experiments on aromatic nitro compounds are incomplete, mainly because of the formation of ill-defined, high-molecular-weight organic compounds. A thorough understanding and more complete characterization of the products of the hydrolysis of TNT and tetryl is still lacking. The complexity of the intermediates may preclude any more exact identification than one based on elemental analysis and functional group identification.

Finding (Pueblo) EH-6. Process parameters and process control strategies (e.g., energetic feed rates, caustic concentration, and reactor temperature) have not yet been characterized in enough detail to ensure a smooth transition to full-scale operation.

Recommendation (Pueblo) EH-1. A bench-scale comparison of the rates of hydrolysis of tetryl and tetrytol should be undertaken before any process for the destruction of tetryl is planned. The rates should not be based only on data from tests with tetrytol.

Recommendation (Pueblo) EH-2. Any post-hydrolysis treatment technology selected for the Assembled Chemical Weapons Assessment program must be capable of accommodating the possible presence of high-molecular-weight organic compounds generated from aromatic nitro compounds.

Recommendation (Pueblo) EH-3. Insofar as possible, the particle size, feed rate, nature of the feedstock (e.g., dry or slurried), and solids loading in the reactor at Holston should be matched with the operating conditions expected at Pueblo to verify the efficacy and safety of the hydrolysis process for energetic materials.

Recommendation (Pueblo) EH-4. Hydrolysis reactions at pilot scale and full scale must be remotely operated.

3

General Atomics Technology Package

DESCRIPTION OF THE PROCESS

The GATS technology package proposed for the Pueblo Chemical Agent Disposal Facility is based on, and very similar to, the design General Atomics originally proposed for the treatment of all assembled weapons at all chemical munitions storage sites in the United States (NRC, 1999). General Atomics is the sole developer of the GATS process, including the designs for all of the munitions-processing and dunnage-processing equipment. The balance of the plant design and site infrastructure was prepared by Parsons Infrastructure and Technology Group.

The Pueblo design incorporates changes based on the Demonstration I tests and other tests conducted during the EDS that coincided with the data-gathering phase of this report (NRC, 2000). The specific GATS technology package evaluated by the committee is for the treatment of the particular mix of mustard-agent-filled munitions stored at Pueblo (i.e., 105-mm M60 projectiles; 155-mm M104 and M110 projectiles; 4.2-inch M2 and M2A1 projectiles; and M2 and M2A1 mortar rounds). These munitions are described in Table 1-2. The full process is designed to treat agent, energetic materials, metal parts (including munitions bodies), and dunnage (e.g., wooden pallets and packing boxes used to store munitions), and nonprocess waste (e.g., plastic DPE suits; the carbon from DPE suit filters and plant HVAC filters; and miscellaneous plant wastes).

The 16 unit operations are shown in Figure 3-1 and discussed below. Figure 3-2 is a block diagram showing the major components of the process. The GATS design anticipates the movement of munitions from storage to the munitions demilitarization building (MDB) using modified ammunition vans (MAVs). Transport will be in two steps: first to the on-site munitions storage building (MSB) and then to the unpack area (UPA) in the MDB.

Disassembly of Munitions (Steps 1 to 4)

Steps 1, 2, 3, and 4 of the GATS process (Figure 3-1) incorporate comparatively minor modifications to existing baseline reverse-assembly procedures. These procedures have been used at the Johnston Atoll and Tooele, Utah, baseline incineration-system disposal facilities, where the Army has accumulated more than 10 years of experience in their operation. During reverse assembly by the projectile mortar demilitarization (PMD) machines, fuzes and whole bursters are removed from the projectiles. The General Atomics design for Pueblo uses two parallel (and redundant) PMD machines to meet specified throughput rates (General Atomics, 2000a).

The bursters removed from the munitions are then sheared to access the energetic materials. The sheared burster parts and fuzes are then transferred to the ERH. The shearing step is a mechanical cutting operation involving shearing equipment used in the baseline disassembly process and in nonchemical weapons applications. All propellant material removed either from the storage containers or from the munitions is fed into the ERH along with the fuzes and sheared bursters.

Hydrolysis of Energetic Materials (Steps 5 and 6)

Energetics Rotary Hydrolyzer (Step 5)

Step 5 of the GATS process is the ERH, a long, steam-jacketed, rotating cylinder with internal spiral flights and lifting flights. Table 3-1 lists design parameters for the ERH (and the projectile rotary hydrolyzer [PRH] discussed in Step 8). The full-scale ERH can be characterized as a series of identical chambers through which materials pass and in which the hydrolysis reaction occurs. The GATS design for Pueblo would use two ERHs. Hot water and NaOH solution, along with energetic materials and associated metal parts from the PMD operation, are fed into the ERH and flow

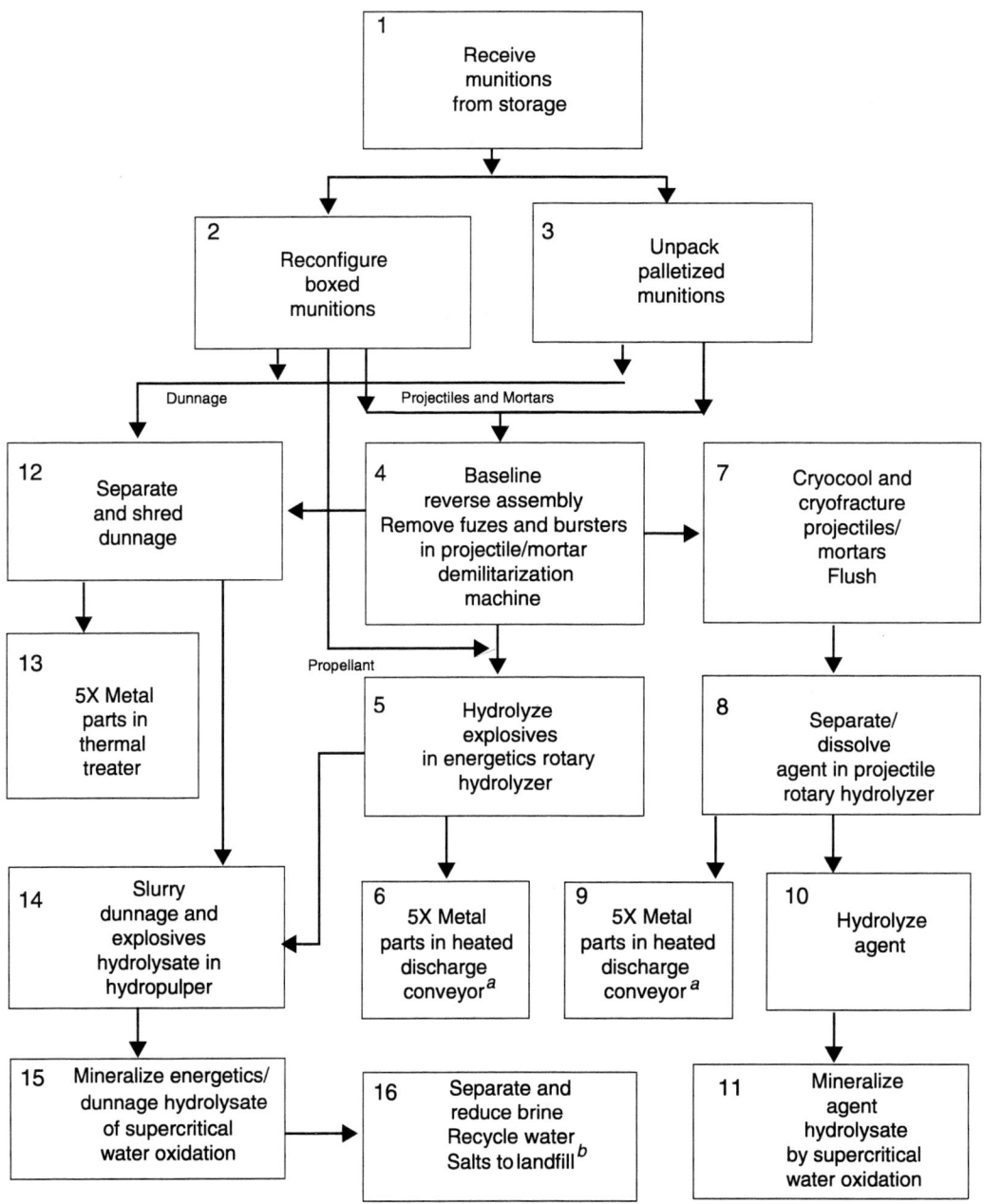

FIGURE 3-1 Simplified schematic flow diagram for GATS. Source: Adapted from General Atomics, 2000a.

concurrently through the ERH. Hydrolysis of the energetic materials by the caustic leaves only small pieces of residual energetics.

The spiral flights in the ERH, which transport material axially along the cylinder, create baffles that divide the cylinder into a series of batches. The lifting flights slowly agitate the hydrolyzing solution with the energetic materials and metal parts. The drum is steam heated on the outside surface to maintain the ERH contents at 100°C to 110°C (212°F to 230°F), which is higher than the melting point of TNT-based energetic materials.

The drum rotates slowly, and each batch moves through the ERH with a residence time of approximately 2 hours. General Atomics claims this time is sufficient for complete

GENERAL ATOMICS TECHNOLOGY PACKAGE

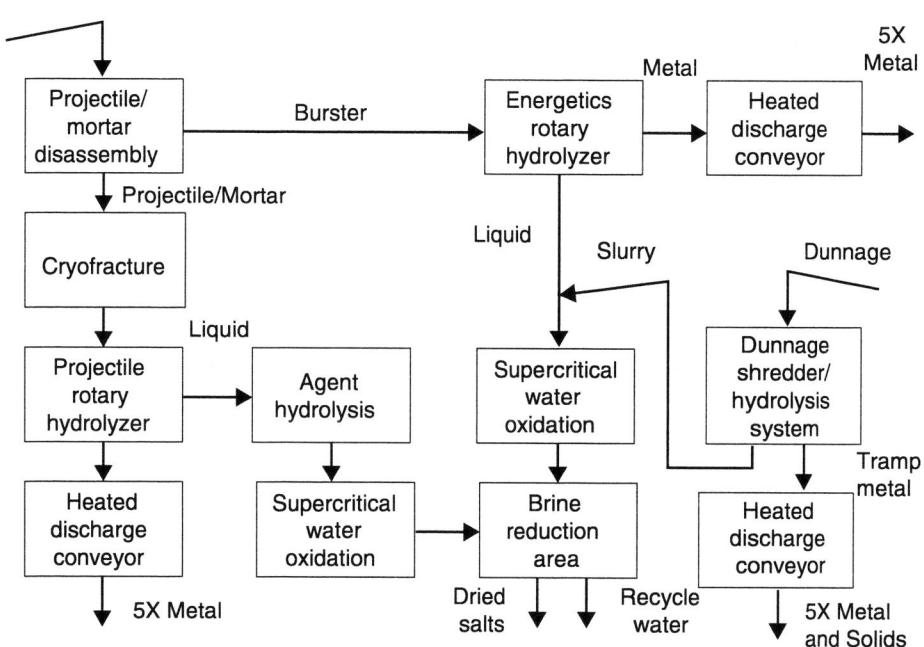

FIGURE 3-2 Simplified block diagram of GATS process components. Source: Adapted from General Atomics, 2000a.

TABLE 3-1 Design Parameters for GATS ERH and PRH

Design Parameter	Energetics Rotary Hydrolyzer (ERH)	Projectile Rotary Hydrolyzer (PRH)
Residence time (nominal) (hr)	2	1
Drum diameter (ft)	4	6
Drum length (ft)	20	40
Flight	Helical, 2.5-ft pitch/ 20 ft long	Helical, 2.5-ft pitch/ 40 ft long
Lifting flights	To be determined	To be determined
Rotations per hr (nominal)	4	9 to 18
Operating temperature	~105°C (221°F)	~100°C (212°F)
Operating pressure	Ambient	Ambient
NaOH solution (12–20%) per pound of energetic (lb)	5.6	0
Process water per pound of munition (lb)	0.5	0.6

Source: Adapted from General Atomics, 2000a.

hydrolysis of the energetic materials in the munitions at Pueblo. Multiple energetic items and their associated materials are fed to the ERH at a rate that results in a maximum explosive loading per flight of about 15 lb TNT equivalent.

At the discharge end of the ERH, the remaining metal parts, which include nonhydrolyzed fuzes and booster cup components, are lifted out of the solution by the spiral flight and fed into a chute leading directly into a heated discharge conveyor (HDC, discussed in Step 6). Immediately upstream, the liquid hydrolysate is separated from metal parts by draining through a perforated section of the ERH drum and accumulated in a sump. The liquid discharge is configured to maintain a minimum depth of 12 inches in the ERH drum. The hydrolysate is pumped to a continuously stirred reactor, where additional hydrolysis can take place if necessary. The NaOH also dissolves any aluminum present in the munitions, converting it to aluminum hydroxide. To prevent clogging of downstream components by the aluminum hydroxide, phosphoric acid is added to form a phosphate precipitate, which is removed by settling or filtration.

Air is drawn through the ERH to remove hydrolysis vapor and fumes, including hydrogen produced from the hydrolysis of aluminum in the munitions. Sufficient air is used to keep the hydrogen concentration well below the lower explosive limit. The air is passed through an air-treatment system consisting of a scrubber, a condenser, and carbon filters and is then exhausted to the plant ventilation system.

Treatment of Metal Parts (Step 6)

In Step 6 of the GATS process, metal parts from the ERH pass through a chute to an electrically heated discharge conveyor (HDC) purged with nitrogen. The metal parts are heated to 1,000°F (538°C) and held at temperature for at least 15 minutes to decontaminate them to a 5X level.[1] The heat causes any residual energetic materials that might remain in the solids (e.g., fuzes) to decompose. The solids leaving the HDC are cooled and disposed of off-site. Off-gases from the HDC are passed through the activated carbon filter system for the explosion containment room (ECR).

Separation of Agent from Munition Bodies and Agent Hydrolysis (Steps 7 to 10)

Cryofracture of Munitions (Step 7)

Step 7 of the GATS process is the projectile agent removal system. After the energetic materials and associated metal parts have been removed and sent to the ERH, the agent cavity of the munition body is accessed by cryofracture and the free agent is drained. Two streams are produced from this step, liquid agent and agent-contaminated metals.

The munitions are first embrittled by cooling in a liquid nitrogen bath (77°K; –321°F; –196°C) and then transferred via an overhead crane to a hydraulic press that cracks open the agent cavity, exposing the frozen, solidified, or gelled agent and agent heels. Any free liquid agent is collected in receivers under the cryofracture press.

The cryocooling bath is modeled after commercial food-freezing tunnels. Key design parameters for the two cryofracture system trains are given in Table 3-2. Projectiles and mortars minus energetic components are conveyed from the PMD ECR to the cryobath loading station in a horizontal orientation. A cryobath loading robot places each round onto a moving link belt that conveys the munitions completely submerged through the liquid nitrogen bath. The residence time in the liquid nitrogen is sufficient to freeze the munition and associated agent to the temperature of liquid nitrogen. At the discharge end, the belt lifts the munitions out of the bath and onto the anvil of the cryopress.

The hydraulic presses (cryopresses) are described in Table 3-2. The press base is a tilt table that discharges the cracked munitions into a discharge chute, which in turn discharges both the metal and frozen agent into the feed chute

TABLE 3-2 Key Design Parameters for GATS Cryofracture Systems (Two Trains)

Design Parameter	Specification
Cryocooling conveyor	
Dimension (ft)	57 (L) × 4 ft (W) × 4 ft (H)
Maximum speed (ft/min)	1.5
Capacity	38 munitions
Liquid N_2 capacity (gal)	3,000
Munition residence time (min)	38 (in bath)
Cryofracture press	
Tonnage	500
Stroke	To be determined
Cycle time (s)	15–30
Ventilation airflow (lb/hr)	4,800
Liquid N_2 usage	1 lb liquid N_2/lb munition + 400 lb/day boil-off
Flush water per fracture (gal)	0.40

Source: Adapted from General Atomics, 2000a.

of the PRH. The present level of design makes no provision for separate removal and treatment of the liquid agent, which now passes into the PRH along with the remaining munitions components.

The cryofracture system was developed and tested by General Atomics for the Army and Air Force under previous programs for cracking solid-fuel rockets and other agent-free munitions (NRC, 1991). Munitions-processing bay-bridge robots, as used in the baseline system, have been fitted with new end effectors for loading and harvesting munitions from the cryobath and transferring them into the press.

Projectile Rotary Hydrolyzer (Step 8)

In Step 8 of the GATS process, accessed frozen agent is hydrolyzed with hot water, and agent-contaminated metal parts from the cryofracturing step are washed in one of two parallel PRHs. The PRHs are larger than the two ERHs (see Table 3-1) but similar to them in function and construction. Each PRH is externally steam heated to maintain the temperature of the metal parts near the boiling point of the water used for hydrolysis.

The drum of the PRH is fitted with a spiral flight and lifting flights to transport and mix the munition fragments axially along the drum from feed to discharge. A stationary shell of thermal insulation encloses the drum and reduces the heat loss to the room. Water introduced at the discharge end flows countercurrent to the flow stream of solids. The liquid level in the PRH is deeper than the height of the spiral flight to ensure that the liquid flows toward the feed end. The hot water, which dissolves the frozen agent and/or agent heels,

---

[1] Treatment of solids to a 5X decontamination level is accomplished by holding a material at 1,000°F for 15 minutes. This treatment results in completely decontaminated material that can be released for general use or sold (e.g., as scrap metal) to the general public in accordance with applicable federal, state, and local regulations.

is discharged through a screen at the feed end of the PRH, separating the solution from the freshly introduced metal fragments. At the discharge end, the spiral flight lifts the metal fragments out of the solution and discharges them through a chute directly onto an HDC, which is distinct from but similar to the HDC to which ERH materials are discharged.

The PRH hydrolysate is discharged to a stirred tank to complete the hydrolysis of agent. Air is drawn through the PRH to remove volatile materials, and the gaseous effluent is passed through a scrubber and carbon filters prior to release to the plant's HVAC system.

Treatment of Metal Parts after PRH Processing (Step 9)

Step 9 of the GATS process is the treatment of the metal parts from the PRH in an HDC, a different unit from the one described in Step 6, but similar in design and function.

Completion of Agent Hydrolysis (Step 10)

In Step 10 of the GATS process, the PRH hydrolysate solution (and any residual agent) is transferred to one of four well-mixed reactor vessels, where caustic solution hydrolyzes the remaining agent and where the hydrolysate is stored pending verification of agent destruction. These reactor vessels are similar in design to the ones that will be used in the hydrolysis of bulk mustard agent at the Aberdeen Chemical Agent Disposal Facility. See Table 3-3 for parameters of the neutralization system tanks.

Treatment of Agent Hydrolysate by Supercritical Water Oxidation (Step 11)

Step 11 of the GATS process is treatment of the agent hydrolysate from the PRH by SCWO. The GATS design uses two SCWO reactors (specifications, including system component sizing assumptions, are shown in Table 3-4). The specifications in Table 3-4 also apply to the two other SCWO reactors in the GATS process treating combined-energetics hydrolysate and slurried dunnage (described in Step 15). The reactors are intended to oxidize organic materials, including the agent hydrolysate, in an aqueous solution of about 10 weight percent hydrolysate. The SCWO reactors operate at approximately 650°C (1,140°F); a nominal operating pressure is approximately 3,400 psig. The conditions are above the critical temperature and pressure of pure water and of the solution. The oxidizer is either ambient air or a synthetic air consisting of a mixture of oxygen and nitrogen at a 21:79 ratio by volume at a feed rate in excess of the stoichiometric requirement. The oxidation reaction is autogenous (i.e., the heat released from the oxidation reaction is sufficient to maintain the reactor temperature). Isopropyl alcohol is added as an auxiliary fuel whenever needed to maintain an autogenous feed mixture to the SCWO reactor.

During system start-up, an electric preheater and the oxidation reaction of isopropyl alcohol are used to heat the reactor to the desired operating temperature. Once the reactor

TABLE 3-3 Key Design Parameters for the GATS Projectile Agent Hydrolysis System

Design Parameter	Specification
Batch preparation tanks	
Number	4
Capacity (gal)	5,000
Materials	TBD
Neutralization tanks	
Number	3
Capacity (gal)	8,000
Materials	TBD
Cycle time (hr)	6
Planned batches/day/tank	2

Source: Adapted from General Atomics, 2000a.

TABLE 3-4 Equipment Sizes for the Full-scale SCWO System[a]

Equipment Component	Requirement
Water feed pump	8 gpm, 3,800 psi
Hydrolysate feed pump[b]	16 gpm, 3,800 psi
Quench pump	52 gpm, 3,800 psi
Auxiliary fuel pump	1 gpm, 3,800 psi
High-pressure oxygen system	6,400 lb/hr
High-pressure nitrogen system	TBD
Hydrolysate tank (including mixer)	10-hr holdup, 25% free space, 12,000 gal
Water tank	4-hr holdup, 25% free space, 2,400 gal
Auxiliary fuel	750-gal tank
Transfer pumps	4 pumps
Start-up preheater	~600 kW
Reactor	12.5 inches (process ID) × 19 ft long
Reactor liner	0.030-inch titanium pipe
Cool-down heat exchanger	3,000 kW
Pressure letdown	Redundant valves
Liquid separation/holdup tanks and quench supply	600 gallons, 50 psi

[a]Two scaled-up SCWO units are required for treatment of agent hydrolysate and for energetics hydrolysate (four trains altogether).
[b]For SCWO units treating agent hydrolysate only. For SCWO units treating energetics hydrolysate, this pump is replaced by the high-pressure slurry feed system with 11 gpm throughput requirement.

Source: Adapted from General Atomics, 2000a.

is at operating temperature, the preheater is turned off, hydrolysate flow is initiated, and auxiliary fuel flow is reduced.

The SCWO reactors are operated cyclically. That is, once started, hydrolysate is oxidized continuously for 22 hours, after which the SCWO reactor is cooled down and flushed at high pressure to clear out any buildup of salts insoluble at SCWO operating conditions. After every three flush cycles, the reactor is cooled and depressurized for weekly maintenance.

The fluid discharged from the SCWO reactor passes through a cooler and enters a phase-separation vessel. Gases and liquids then flow to separate pressure-reduction stations before entering a low-pressure phase-separation vessel. Noncondensable gases, mostly carbon dioxide, are monitored and filtered before release to the environment via the plant HVAC system. Liquids are monitored and transferred to the brine recovery area, which is identical to the one used in the baseline incineration system. If fluid does not meet release specifications, it is returned to a storage tank for off-specification product and reprocessed in the SCWO reactors.

The GATS SCWO reactor design includes a replaceable titanium inner liner surrounded by a gap purged with nitrogen. With this design, the liner can be periodically replaced to compensate for corrosion, and the reactor shell can be operated at a lower temperature. The nitrogen purge should keep water from contacting the shell material. The liner is made of commercially available Grade 2 titanium. The results of the 500-hr test and other tests are discussed later in the chapter.

Processing and Treatment of Dunnage and Energetics Hydrolysate (Steps 12 to 15)

Steps 12 through 15 address the processing and treatment of dunnage and nonprocess wastes in the DSHS. After shredding and micronization, these materials are combined with energetics hydrolysate from the ERH for subsequent treatment by SCWO.

Step 12 of the GATS process is the separation, shredding, and grinding of dunnage and nonprocess wastes. The design parameters for the DSHS are listed in Table 3-5. The technology provider assumes that all dunnage and nonprocess wastes are contaminated with agent. The organic materials (e.g., wood, paper, rubber, plastic, metal-free DPE suit material, and spent carbon) are reduced in a series of steps to a particle size of less than approximately 1 mm and fed to two commercial hydropulpers; a grinding pump then transfers the slurried material to the high-pressure pumps that feed it to the SCWO reactor.

Wood dunnage (e.g., pallets and boxes) is reduced in a dedicated low-speed shredder, hammer mill, and micronizer to produce a fine wood flour suitable for slurrying. Metal-scavenging magnets are used at various points in the process to remove tramp metal, such as shredded nails.

TABLE 3-5 Design Parameters for the GATS DSHS

Material	Specification
Wood	
Size-reduction rate (lb/hr)	1,650
Particle size (mm)	<1
Plastic/rubber	
Size-reduction rate (lb/hr)	70
Particle size (mm)	<1
Spent carbon	
Size-reduction rate (lb/hr)	30
Particle size (mm)	<0.5

Source: Adapted from General Atomics, 2000a.

Metal parts are manually cut and removed from each DPE suit at the same time the worker is cut out of the suit upon exiting the Level 1 area. The metal fittings are sent to the PRH for treatment. The metal-free DPE suit material is fed to a two-stage size-reduction system. In the first step, the material is shredded in a dedicated granulator. Tests showed that further mechanical size reduction in the granulator was ineffective because the polyvinyl chloride suit material melts rather than tears. Therefore, after shredding, the DPE suit material is cryocooled in a small nitrogen bath and fed to a micronizer. Several passes of the cryocooling and micronizing steps may be necessary to ensure adequate size reduction.

Spent activated carbon from plant HVAC filters is ground wet in a dedicated colloid mill. The wood, plastic and rubber, and spent carbon materials are thus processed in three separate equipment lines. The resulting dry material is placed in storage bins prior to introduction into the hydropulpers, described in Step 14.

Step 13 of the GATS process is the thermal treatment and decontamination of metal parts from dunnage and DPE suits to a 5X level. The original design used a separate metal treating furnace. In the current design, the metal parts are sent to one of the HDCs.

Step 14 is the slurrying of the dry, size-reduced dunnage materials and nonprocess wastes from Step 12 with energetics hydrolysate in preparation for feeding to one of two dedicated SCWO reactors. The hydrolysate fluid from the ERH is pumped into a holding tank, where phosphoric acid is added to precipitate aluminum. The hydrolysate fluid is then filtered to remove the precipitated aluminum and transferred to one of the two hydropulper tanks. Spent decontamination solution used in various decontamination operations in the plant also goes to the hydropulper tanks. Additional water or a dilute solution of NaOH is added as needed to adjust water content, neutralize any residual agent, and otherwise adjust the slurry to meet the feed chemistry requirements of the SCWO reactors. Other additives are used to ensure that the solids remain in suspension and that the slurry can be reliably pumped and processed in the SCWO reactor system.

In the hydropulper tanks, the ERH hydrolysate fluid is mixed with the dry material produced from shredding and micronizing the organic dunnage and other waste materials to produce a slurry. The hydropulper tanks are continuously stirred and periodically sampled prior to the transfer of their contents to the hydrolysate storage tank, from which the slurried dunnage and neutralized energetics are pumped to the two SCWO reactors.

Step 15 of the GATS process is the treatment of the slurried dunnage and neutralized energetics in one of two dedicated SCWO reactors. These reactors are different units from those described in Step 11 but are similar in design and function.

Water Recovery and Salt Disposal (Step 16)

Step 16 is concentration of the brine from the four SCWO reactors using evaporation/crystallization equipment to reclaim the water and generate solid salt cake for off-site disposal. No specific design details for brine water recovery and salt cake disposal had been developed for this operation at the time this report was prepared. The evaporation/crystallization step has been eliminated from the designs of the facilities at the Newport and Aberdeen sites, and brine at Tooele is being sent off site rather than processed in the facility's brine-reduction area. It is the committee's understanding that off-site disposal of the SCWO liquid effluent, instead of concentrating the brine, is being investigated for the Pueblo site.

INFORMATION USED IN THE DEVELOPMENT OF THE ASSESSMENT

To produce a report by July 2001, the committee had to review test data and documented design developments concurrently with ongoing EDS activities. Thus, most of the data were available only in draft form. Some information was provided to the committee orally through briefings by the PMACWA and by the technology providers. In developing its assessment, the committee used several sources of information:

- General Atomics EDP (General Atomics, 2000a)
- General Atomics EDS study plan (General Atomics, 2000b)
- General Atomics EDS draft final report (including preliminary results of tests in progress at the time this report was prepared) (General Atomics, 2001)
- design-review meetings (General Atomics, 2000c, 2000d, 2000e, 2000f, 2000g)
- briefings by PMACWA personnel (Susman, 2000a, 2000b)
- visits to EDS test sites at Dugway Proving Ground

Engineering Design Package

The engineering design drawings and associated documentation for the proposed facility provided to the committee were very extensive. They were complemented by the information committee members received from design-review meetings where the technology provider and other contractors associated with the General Atomics EDS discussed and clarified intermediate plans and activities associated with the development of the facility design (General Atomics, 2000c, 2000d, 2000e, 2000f). The committee concentrated its efforts on evaluating the following critical components of the GATS process design that had been identified as potential concerns in the two reports by the ACW I Committee (NRC, 1999; NRC, 2000):

- the long-term reliability of the SCWO system
- the advantages and disadvantages of cryofracture over baseline technology as a means of accessing the agent in the munitions
- the ability of the rotary hydrolyzers (both the PRH and the ERH) to process their respective feed materials in a reasonable time and with acceptable safety and reliability

Engineering Design Studies Tests

The GATS EDS tests were designed to provide data for the evaluation of the ERH, DSHS, and SCWO components of the GATS process (DOD, 2000). Demonstration I tests conducted in 1999 were designed to address the issues of SCWO reactor design and reliability (General Atomics, 1999a). Although the scope and schedule of those tests were not sufficient to address these issues fully, specific problems were identified that had to be addressed during the EDS-phase testing. The Demonstration I results were evaluated by the NRC and published in a supplemental report (NRC, 2000); key results are summarized in the "Previous Findings and Recommendations" section at the end of this chapter. The results were used by the Army and its contractors in the design of tests included in the General Atomics EDS. The EDS tests evaluated three components of the GATS: the ERH, the SCWO system, and the DSHS.

ERH Testing

Specific test objectives for the EDS ERH tests were as follows (General Atomics, 2001):

- Demonstrate the effects on the hydrolysis of M28 propellant of the following changes in operating procedures (relative to results of the PMACWA Demonstration I test program):
  — rocket motor lengths shorter than 12 inches (4 inches and 8 inches)

— multiple pieces of rocket motor lengths (4 inches)
— NaOH concentrations of 12 to 14 M (molar concentration)
— temperatures to 120°C (248°F)
- Results should be compared with data from the PMACWA Demonstration I test program on 2-inch lengths with steel casing in 12 M NaOH at 110°C (230°F).
- Demonstrate containment of fugitive emissions and characterize gas, liquid, and solid process streams from the ERH process to supplement data generated during the PMACWA Demonstration I test program.
- Demonstrate the effect of higher caustic concentration and bath temperature (19 M [50 weight-percent] NaOH, 136°C [277°F] or highest allowable) on the rate of hydrolysis of M28 propellant.
- Confirm that nitrosodiphenylamine and other reaction products were removed from the ERH off-gas by a negative-draft pollution-abatement system (a condensing scrubber and downstream filters).

The ERH tests consisted of rate-of-hydrolysis tests for the following feed materials:

- single 4-inch rocket motor segments in 12 M NaOH solution at 110°C (230°F)
- single 8-inch rocket motor segments in 12 M NaOH solution at 110°C (230°F)
- multiple rocket motor segments in 12 M NaOH solution at 110°C (230°F)
- single and multiple rocket motor segments in 12 to 14 M NaOH solution at 120°C (248°F)

The results of the ERH tests were as follows (General Atomics, 2001):

- M28 propellant can be hydrolyzed from single 4-inch rocket motor segments at 110°C (230°F) for 5 hours with <5 g propellant remaining.
- M28 propellant can be hydrolyzed from single 8-inch rocket motor segments at 110°C (230°F) for 5.5 hours with <5 g propellant remaining.
- M28 propellant can be hydrolyzed from as many as eight 4-inch rocket motor segments (one complete rocket) at 110°C (230°F) for 7 hours with <30 g propellant remaining, or for 7.5 hours with <5 g propellant remaining.
- M28 propellant can be hydrolyzed from as many as eight 4-inch rocket motor segments (one complete rocket) at 120°C (248°F) for 3 hours with <30 g propellant remaining, or for 3.5 hours with no propellant remaining.
- The negative-draft pollution-abatement system with condensing scrubber and filters effectively captured nitrosodiphenylamine and other fugitive emissions.

The ERH test unit was operated in a batch mode rather than a continuous mode. Results from the EDS tests are to be used for sizing the full-scale ERH and establishing the residence times of the munitions in the hydrolysis solution. The testing further investigated the effect of hydrolysis parameters, specifically caustic concentration and process temperature. Although no rockets are stored at Pueblo, ERH tests with sections of rocket motors were done to optimize the processing conditions for M28 propellant. Test data were not generated during the EDS for the propellants associated with the 4.2-inch mortars or with the 155-mm and 105-mm projectiles stored at Pueblo. The PMAWCA-sponsored EDS program on hydrolysis of energetic materials was designed to address these materials (discussed in Chapter 2), and the results of that program are applicable to the General Atomics design.

SCWO Testing

The GATS SCWO system is designed to oxidize an aqueous organic feed to carbon dioxide, water, and salts. The EDS testing of the SCWO system had the following objectives (General Atomics, 2000b):

- Demonstrate long-term continuous operability without plugging.
- Demonstrate acceptable corrosion rate.
- Demonstrate that any feed additives for control of salt transport do not interact with feed and/or equipment to generate salt plugs or accelerate corrosion.
- Determine a maintenance schedule and the frequency of shutdowns based on the results of this long-term testing.
- Generate data for use in validating the development of a SCWO model sponsored by the Army Research Office.

Two types of test runs were performed during the GATS EDS for SCWO: work-up tests and 500-hr duration tests. The work-up tests were trials to verify system upgrades and operating conditions for simulant and hydrolysate processing. These tests were performed at varying conditions to determine the best conditions for simulant destruction and salts or solids transport. Starting with the flow rates used in the Demonstration I testing, the goal was to increase flow to provide residence times consistent with those anticipated for the full-scale plants.

Flow rates for the EDS tests were finalized based on the results of the work-up trials. The primary goal of the 500-hr EDS tests was to demonstrate long-term operability of the system. Table 3-6 shows the original plans for the EDS tests. In addition to those tests, limited tests with actual mustard agent and nerve agent GB hydrolysates were also scheduled. Many of the planned tests were not performed, however, because of operational and equipment problems with the

TABLE 3-6 Feeds and Duration of Planned SCWO Tests

Feed	Duration (hr)
Agent hydrolysate	
Mustard agent HD hydrolysate (15 wt % HD mixture)	12
Mustard agent HD hydrolysate simulant	488
Nerve agent GB hydrolysate	12
Nerve agent GB hydrolysate simulant	488
Energetics hydrolysate/dunnage	
Composition B hydrolysate/M28 hydrolysate/ aluminum hydroxide/dunnage with PCBs	11–20
Composition B hydrolysate/M28 hydrolysate/ aluminum hydroxide/dunnage	500
Tetrytol hydrolysate/aluminum hydroxide/dunnage	500
GB hydrolysate simulant for Army Research Office model	10

Source: Adapted from General Atomics, 2000a.

SCWO system. Test results that were available are discussed below.

The EDS tests evaluated a number of different liners for the reactor interior. Initially, General Atomics had proposed using a platinum liner for Demonstration I testing. However, because a platinum liner could not be fabricated on the Demonstration I schedule, General Atomics used a Hastelloy reactor with no liner during the Demonstration I testing. Serious corrosion of the SCWO reactor resulted, a problem that had been identified previously (NRC, 2000).

On the basis of preliminary EDS testing, General Atomics determined that platinum might be an appropriate liner material to treat both nerve agent VX and mustard agent HD hydrolysate. As a result, the first EDS SCWO tests with HD hydrolysate were conducted in a reactor with a platinum liner. The platinum liner did not perform adequately. Degradation was first observed in the form of blisters after 138 hot hours.[2] Collapse and regrowth of blisters presumably led to a breach of the liner, observed at 170 hot hours. The liner continued to serve as a reasonably good corrosion barrier for another 100 hours of feed exposure, for a total of 315 hot hours. But by then, the poor condition of the liner necessitated that it be retired from service. Analysis of the SCWO effluent showed adequate destruction (99.9999 percent) of organic materials in the feed, but compounds of platinum were also identified.

Because the performance of platinum in the HD hydrolysate-type environment was based on limited preliminary data, a contingency liner of titanium had also been fabricated for the EDS tests. Shortly after the failure of the platinum liner, this Grade 7 titanium (0.5 percent palladium) liner was installed in the reactor. The titanium was 0.030-inch thick and had an outer sleeve of Hastelloy C-276, similar to the sleeve used with the platinum liner. The primary difference in liner geometry was the absence of bimetallic welds between the Hastelloy sleeve and the titanium liner because titanium and Hastelloy C-276 are not compatible materials for welding. The titanium liner originally hung from a ring welded around its top edge; in later designs, it rests on several lugs welded to the bottom ID of the C-276 sleeve or on the discharge cone of the reactor. No seal isolates the space between the titanium and Hastelloy C-276 from the process fluid. A slow stream of nitrogen is passed between the titanium liner and the Hastelloy reactor shell to cool the shell and prevent contact with process fluid.

Table 3-7 is a summary of the operating log from the first round of EDS tests of the SCWO system using two types of titanium liners. The first titanium liner failed after 33 hot hours and was replaced with a second, identical liner, which failed after 161 hot hours. Meanwhile, the first liner was patched and reinserted. It was operated for 102 hours. On November 14, 2000, the patched (original) titanium liner was replaced with a new one manufactured from titanium pipe rather than custom rolled from titanium sheet. The new liner was of a different grade and thickness of titanium from the original one. Titanium pipe is a commercially available product that, according to General Atomics, is far less expensive and easier to produce than the original rolled-titanium liner (Spritzer, 2000a; Hong, 2001). The reactor was operated with the new liner for 115 hours, when corrosion at the top exceeded 50 mil, approximately one-half the liner thickness. The liner was then inverted in the reactor (with the undamaged portion placed at the top) and operated for an additional 106 hours, when corrosion at the top reached 85 mil. The liner was inverted and operated again for 332 hours. No corrosion results were presented for the liner after this period of usage (General Atomics, 2001).

On the basis of these results, a 500-hr mustard-agent hydrolysate run was approved by the PMACWA. The following criteria for success were established for the 500-hr run (General Atomics, 2001):

- *Corrosion.* Liner wear-surface change (liner flip or replacement) not more frequent than every 66 hours of HD hydrolysate feed, on average.
- *Salt transport.* Salt flush not more frequent than every 22 hours of HD hydrolysate feed, on average.
- *Feed composition.* Demonstrate control of feed composition to maintain corrosion and transport of salts within the bounds defined by the preceding criteria. Real-time indicators of satisfactory feed composition include reactor temperature and pressure profiles and effluent pH and turbidity.
- *Organic destruction.* Liquid effluent total organic carbon (TOC) <10 mg/L, on average.
- *Availability.* Process feed at least 39 percent of calen-

---

[2]Hot hours are defined as operating hours whether on supplementary fuel (isopropyl alcohol), hydrolysate, or hydrolysate simulant.

TABLE 3-7 Corrosion of Titanium Liners During GATS EDS Work-up Tests

Date	Event	Fuel Hours	Feed Hours	Total Hot Hours
10/4/00	Start Ti rolled-sheet liner #1 (Grade 7, 0.030 inches thick).	0	0	0
10/5/00	Small hole in liner about 2 inches below nozzle tip. Two patches each several inches square with wall thinning of about 10 mil.	10	23	33
11/1/00	Start Ti rolled-sheet liner #2 (Grade 7, 0.030 inches thick).	0	0	0
11/2/00	Ti pit depth ~10 mil. General thickness loss <~1 mil.	6	33	39
11/12/00	~2,000 small pits counted with maximum depth ~10 mil. Switch to new liner for 5% HD hydrolysate simulant tests.	16	145	161
11/14/00	Start repaired Ti rolled sheet liner #1.	10	23	33
11/17/00	Switch to testing of Grade 2 pipe liners.	17	85	102
11/18/00	Start Ti pipe liner #1 (Grade 2, 0.110 inches thick).	0	0	0
11/19/00	Pitting noted 16 inches below nozzle tip. Maximum depth ~10 mil.	2	20	22
11/22/00	Pitting from 6 inches to 18 inches below nozzle tip. Maximum depth ~20 mil. Maximum general corrosion ~5 mil.	7	47	54
11/30/00	Maximum general corrosion ~50 mil primarily in top 3.5 inches below the nozzle tip. Maximum pit depth ~20 mil.	10	105	115
12/1/00	Start inverted Ti pipe liner #1 (Grade 2, 0.110 inches thick).	10	105	115
12/5/00	Maximum corrosion ~85 mil, a bit less than 1 mil/hr. Corrosion primarily in top 3.5 inches below the nozzle tip.	21	200	221
12/12/00	Ti pipe liner #1 returned to original orientation.	27	237	264
12/17/00	Ti pipe liner #1 broken during removal from Hastelloy sleeve.	38	294	332

Source: General Atomics, 2001.

dar time, with minimum outages, excluding events not directly relevant to Pueblo operations.

- *Run time.* The 500-hr run clock begins when steady state is attained with full hydrolysate flow. The 500-hr run time applies to normal system operation and therefore includes planned system flush cycles. Unplanned shutdowns and recovery from unplanned shutdowns back to steady-state operation are not counted in the 500-hr run time.

The ACW II Committee was briefed on these tests on February 8, 2001, and a report was provided to the NRC shortly thereafter (General Atomics, 2001; Hong, 2001). The log for the 500-hr tests is reproduced in Appendix B. The results are discussed as part of the committee's evaluation of the GATS technology package.

Engineering-scale Tests of SCWO for the Newport Site

The Project Manager for Alternative Technologies and Approaches in the CSDP has already established an engineering-scale test (EST) program to support the planned use of SCWO to treat nerve agent VX hydrolysate that will be produced during disposal operations at the Newport site, where only bulk VX is stored. The EST uses a one-tenth-scale General Atomics SCWO reactor design and is being conducted at GDS, Inc., Corpus Christi, Texas. Although VX is not stored at Pueblo, and the EST results for treatment of VX hydrolysate are not directly applicable to the use of SCWO in the GATS design for Pueblo, these test results were useful for the overall evaluation of the operability of the GATS SCWO system. The following test description is taken from a letter report prepared by another NRC committee (NRC, 2001):

> The EST facility in Corpus Christi, Texas—designed, constructed, and presently operated by General Atomics—is a one-tenth-scale pilot version of the SCWO reactor planned for NECDF (Newport Chemical Agent Disposal Facility). The EST facility was initially scheduled to produce the data necessary to confirm the SCWO reactor system design and operational parameters by April 1999. However, difficulties have arisen, including problems in (1) fabrication (flange welds have failed), (2) design (two multilayered removable platinum liners have proved to be unworkable), (3) materials of construction (the platinum liner has developed a bulge and separated from its supporting structure, and platinum has migrated into a deposit of solids), and (4) operation (erosion has been found in the feed nozzle and the automatic control system, and the high-pressure oxygen supply system failed in one test). EST processing of surrogate hydrolysate solutions began in late October 1999, but results to date have not provided sufficient data to support scale-up.

DSHS Testing

The GATS DSHS is used to reduce dunnage to a shredded product that can be slurried with energetics hydrolysate from the ERH in a hydropulper for subsequent processing by SCWO. EDS testing of the DSHS was designed to meet the following specific objectives (General Atomics, 2000b):

- Demonstrate all changes (relative to the PMACWA Demonstration I tests) to the dunnage-shredding equipment proposed for the full-scale design and verify improved efficiency and uninterrupted operation (e.g., avoidance of nesting and unit overloads) and a particle size of <1 mm for wood and plastic/rubber and <0.5 mm for carbon.
- Generate information required for design of the dust/agent vapor-emission control system.
- Verify size reduction for carbon in the carbon grinder sufficient for downstream SCWO processing.
- Verify feasibility of removing DPE metal parts fixtures for full-scale facility operation.

The tests on DPE suit material and wood were planned to address size-reduction and material-transport problems identified during the Demonstration I testing (NRC, 2000).

Since the technology required for EDS testing had been developed for other applications in industry, full-size commercial equipment was used in the EDS testing, rather than the smaller size proposed for Pueblo. The particle size for wood dunnage was continuously reduced through (1) a dedicated low-speed shredder, (2) a reducing screw feeder, (3) a hammer mill, and (4) a micronizer to produce fine wood flour. DPE suit and butyl rubber simulant materials were shredded in a dedicated granulator, cryogenically cooled in a cryocooler with internal screw conveyor, and reduced in a cryocooled hammer mill. No materials contaminated with agent were tested.

EDS tests for processing of the wood dunnage and for the granulation of DPE suit material were performed using existing General Atomics process equipment (low-speed shredder, hammer mill, micronizer, and granulator) and a wood screw feeder that had been installed during EDS testing at Dugway Proving Ground, Utah. General Atomics also provided guidelines for sampling, operator training, systemization protocols, standard operating procedures, and test plans. The PMACWA provided the operating personnel and arranged for testing support services.

General Atomics' engineers and field technicians specified the appropriate operating conditions to reduce wood particles to less than 1 mm at a process rate of 1,250 lb/hr and DPE suit materials to less than 1 mm at a process rate of 70 lb/hr. Material from the wood pallets processed through the wood process line and DPE material processed in the granulator system were sampled and analyzed to confirm the size reduction of feed materials and the sizing of the overall system.

EDS testing of cryogenic micronization of DPE material was completed at Pulva Corporation facilities in Saxonburg, Pennsylvania. General Atomics provided PMACWA-supplied feed materials, guidelines for sampling, a test plan, and operating procedures. Pulva Corporation provided the test facilities, cryogenic equipment, operating personnel, utilities, sieves and sieve stack agitator, equipment cleanout, and product transport to Dugway Proving Ground. Pulva engineers specified the appropriate operating conditions to reduce rough-granulated DPE suit material to less than 1 mm at a process rate of 70 lb/hr.

EDS testing for micronization of granulated activated carbon was completed at Bematek Systems facilities in Beverly, Massachusetts. General Atomics provided PMACWA-supplied feed materials, guidelines for sampling, carbon slurry-drying techniques, a test plan, and operating procedures. Bematek provided the facilities, wet-milling process equipment, operating personnel, utilities, sieves, sieve stack agitator, equipment cleanout, and product transport to Dugway Proving Ground. Bematek engineers specified the appropriate operating conditions to reduce the size of spent granulated activated carbon to less than 0.5 mm at a process rate of 30 lb/hr.

In summary, in spite of a few minor operating problems, the tests appeared to have been successful. All materials, pallets, carbon, and DPE suit material were reduced to within the size specifications for feeding to the SCWO system; the metal removal devices appear to have performed well and fugitive dust appears to have been controlled. The size reduction of the DPE suit material was of special interest because the technology for heavy polymeric composites is comparatively new.

The work-up and EDS granulation testing demonstrated that DPE suit material could be successfully granulated to less than 10 mm in General Atomics' existing granulator at Dugway Proving Ground. The granulated DPE suit material was then shipped to the Pulva Corporation facilities, where it was successfully size-reduced in Pulva's cryogenic micronization system. Approximately 177 lb DPE suit material was micronized during the first six test runs.

Agent Hydrolysis Studies

Laboratory testing of agent hydrolysis conducted at the Edgewood Chemical and Biological Center (Maryland) led to EDS tests using 15 weight percent mustard hydrolysate for feed to the SCWO reactor.

Materials of Construction Studies

In the course of its information-gathering activities, the committee examined two earlier reports on materials of construction for SCWO reactors that contained pertinent test data applicable to the treatment of HD hydrolysate (General Atomics, 1997; SWEC, 1996).

ASSESSMENT OF PROCESS COMPONENT DESIGN

With the exception of the PMD machines and the SCWO system, all of the components in the unit operations of the GATS process are commercially available. In general, however, this equipment has not been used in applications as demanding as chemical demilitarization. For example, the dunnage-shredding train must produce a metal-free, fine powder from essentially scrap wood. Even though wood shredding and hydropulping are common processes in the papermaking industry, a papermaking machine is far more tolerant of the presence of a few larger particles of wood or a few metal shards than the feed nozzles of the SCWO reactors. A significant issue is the ability of the DSHS operation for wood dunnage, the DPE suit micronizing operation, the slurrying operation, and upstream operations (e.g., the PMD machines and cryofracture system) to produce a consistent feed of the quality required to avoid unscheduled maintenance of the SCWO reactors.

The most significant considerations unique to the GATS process are listed below:

- scale-up of the ERH to a full-scale, continuous-flow system based on test data obtained on a batch ERH test unit module
- shredding of dunnage (e.g., pallets, DPE suits, and gloves) to less than 1 mm by a combination of hammer milling, shredding, and cryogenic milling (for DPE suits and other polymeric waste) with high-shear blending and hydropulping to create a slurry that can be fed into the SCWO reactor for oxidation
- SCWO treatment of agent hydrolysate, energetics hydrolysates, and slurried dunnage

Prior to EDS testing, the ACW I Committee had identified the following issues as critical to the demonstration of the overall GATS process (NRC, 1999, 2000):

- the ability of the SCWO reactor to operate continuously for reasonable periods of time[3]
- the ability of the SCWO units to process shredded dunnage, including shards of metal (e.g., nails in pallets and pieces of the metal connectors in DPE suits) that may elude separation prior to becoming part of the SCWO feed stream
- the ability of the shredding system to achieve ≤1 mm size and to keep tramp metal out of the final SCWO feed
- the ability of the ERH to treat the burster, fuze, and propellant safely and effectively to create a hydrolysate that can be processed in the SCWO system
- the quantity and composition of discharges from the process to the environment via the air, wastewaters, and solid waste

These issues are addressed below in the course assessing the steps in the GATS process.

Disassembly of Munitions (Steps 1 to 6)

The Army has accumulated years of experience with the PMD machinery, which is part of the baseline incineration system at two operating disposal facilities. Although these machines experienced a number of operational problems in the past, they appear to have matured and are an acceptable method of separating energetic components from assembled chemical weapons. In the baseline system, however, PMD machines are used to prepare munitions for incineration. The GATS process (and other ACWA technologies being investigated) propose using them in slightly different ways. The GATS PMD machine is similar to the baseline version with respect to the **Steps 1 to 3** for removing fuzes, bursters, and miscellaneous parts. The discharge/output components for transporting the disassembled parts to their destinations are different, but the changes can be accommodated by well-established engineering methods (e.g., energetics may be cut into smaller lengths). **Step 4,** removal of encased energetic materials by shearing, is a routine operation. In the committee's opinion, shearing in the GATS process at Pueblo should not pose any difficulties beyond the safety issues normally encountered in the handling of energetic materials.

Hydrolysis of Energetic Materials

Step 5, the ERH, is unique to the GATS process. To the best of the committee's knowledge, a system such as this has never been used to hydrolyze solid energetic materials. Although the ERH should prove to be workable, a number of engineering issues will have to be addressed before it can be used in a full-scale disposal facility at Pueblo. The issues still pending following EDS testing are discussed below.

First, the committee is concerned that processing agent-contaminated energetic materials would require verification that no agent is present in the hydrolysate that leaves the ERH. If the complex chemical soup constituting the energetics hydrolysate interferes with analysis for agent, downstream operations (including the high-pressure SCWO) would have to be operated in a Category A environment rather than the planned Category C environment. This would undoubtedly increase the complexity of the SCWO system and pose additional operating and maintenance challenges.

Second, the committee believes ERH testing conducted with a batch, single-chamber flight reactor can adequately simulate the kinetics of energetics hydrolysis in the multi-

---

[3]The committee recognizes the term "reasonable" in this context to mean that required system maintenance does not interfere with an acceptable level of throughput. Doubts had been expressed about the ability of the SCWO reactor to withstand corrosion and to avoid plugging by precipitates and corrosion products.

chamber cascading system configuration proposed for the full-scale ERH, but it cannot simulate the mechanical behavior of the overall system. For example, if the energetic materials contain a granular component that does not hydrolyze, the granular components could accumulate behind the flights of the initial chambers in a full-scale, cascading ERH. According to General Atomics, the ERH design should ensure that loose solids (e.g., small metal parts or cuttings) move forward and do not accumulate (Spritzer, 2000b). However, a gummy/sticky reaction product from an unknown energetic component could cause problems that would require a shutdown and removal by personnel. This situation (or any other maintenance inside the ERH) would pose a severe challenge to maintenance personnel. Enough energetic material could even accumulate to create a hazardous condition. Before building a full-scale ERH, the committee believes it would be prudent to test the continuous (flow, not batch) ERH design to determine (1) its flow characteristics, (2) that energetics hydrolysis is complete, and (3) that no hazardous residues accumulate on the ERH walls.

In summary, EDS testing of the ERH and other testing discussed in Chapter 2 suggest that, although the hydrolysis of energetic materials can be achieved, the hydrolysis process must be further optimized. Fuzes and bursters were successfully hydrolyzed during Demonstration I, and EDS testing of the ERH focused on 4-inch and 8-inch chunks of M28 propellant, which is not among the energetic materials associated with the munitions stored at Pueblo. Nevertheless, tests of energetics hydrolysis in other locations suggest that the larger chunks of M28 are more likely to represent a worst-case condition than the smaller grain of other energetic materials. The results of testing to date indicate that the ERH is a promising technology, but its use with the specific type and configuration of energetic components should be demonstrated. In addition, a continuous version of the ERH should be tested before a full-scale system is built.

Step 6 of the GATS process is the heating of metal parts from the ERH to 1,000°F (538°C) and holding them at temperature for 15 minutes by the HDC to decontaminate them to the 5X level. The committee did not identify any difficulties in this step.

Separation of Agent from Munition Bodies and Agent Hydrolysis (Step 7)

The committee did have a number of concerns about **Step 7** of the GATS process, the projectile-agent removal system, in which the agent cavity is accessed by cryofracture. The EDS testing did not include tests of cryofracture as a means of accessing the agent cavity in projectiles and mortars. After evaluating the design of the cryofracture system, the committee concluded that the potential advantages of this technique over traditional means of accessing the agent (i.e., mechanical shearing or punching) have not been demonstrated. Cryofracture has been successfully demonstrated as a means of breaking up nonchemical projectiles but has not been tested as a method of accessing the agent cavity in projectiles and mortars. This raises two primary concerns regarding the GATS cryofracture operation:

- An unexpected component of a projectile or mortar might incorporate a material that cannot be made brittle at the temperature of liquid nitrogen.
- Solidified agent and other contaminants could accumulate in the cryobath tank that might be released to the surroundings when the bath is drained and the liquid nitrogen evaporated (e.g., during shutdown or during cleaning or other maintenance).

The first concern is based on the behavior of alloys of aluminum, which do not become brittle when cooled to the temperatures of liquid nitrogen (NRC, 1991). To date, no projectiles or mortars have been found to contain aluminum components, so this is only a hypothetical concern for the Pueblo facility.

The second concern relates to safety. Quantities of ice, dirt, and solidified impurities will accumulate in the liquid nitrogen cryobath tank. The committee also believes that some of the munitions might crack from the thermal shock of immersion in the liquid nitrogen bath. Although the agent thus released would quickly freeze in the crack, making the process self-healing (Spritzer, 2000a), some frozen agent could still accumulate in the cryobath tank.

During shutdown of the cryofracture system for maintenance, the liquid nitrogen would typically be drained and the cryobath tank allowed to warm to room temperature. As the tank warms, agent would vaporize and create airborne hazards, which would require that personnel use higher-level protective gear. General Atomics has responded that sufficient decontamination fluid to neutralize any agent could be introduced into the tank as the liquid nitrogen evaporates (Spritzer, 2000b). Appropriate protective gear would be used during the decontamination.

Although cryofracture may offer some performance advantages over mechanical access to the agent cavity of projectiles and mortars, the information available is insufficient to determine if the advantages would be offset by safety concerns and additional maintenance requirements. The committee is concerned because sufficient thought did not appear to have been given to these potential issues. For example, the response to the committee's query on decontamination of any agent in the cryotank ignores the obvious fact that all known decontamination fluids are solids at the temperature of liquid nitrogen. Cryofracture should, therefore, be demonstrated in a way that addresses these concerns before it is implemented. The committee concluded that a decision to use this alternative to mechanical shearing should be governed by considerations of safety, cost, and reliability of the equipment.

Agent Hydrolysis and Metal Parts Treatment (Steps 8 to 10)

In **Step 8** of the GATS process, accessed frozen agent is hydrolyzed, and agent-contaminated metal parts from the cryofracturing step are decontaminated to a 3X level by washing in one of two PRHs. The PRH is similar in design to the ERH, and its application for hydrolysis of agent and munition body fragments from the cryofracturing process appears to be reasonable. As stated in prior NRC reports, hydrolysis appears to be a well-established technique (NRC, 1999, 2000). In essence, the committee believes the PRH is an adequate mixing system to effect hydrolysis of agent and decontamination washing of the metal munition parts.

Step 9 of the GATS process is the treatment of the metal parts from the PRH in an HDC to a 5X decontamination level. Although this is a different unit from the one described for 5X decontamination of metal parts from the ERH, it is similar in design and function, and all comments apply to both units.

Step 10 of the GATS process is completion of hydrolysis of the liquid agent remaining in the hydrolysis solution that drains from the PRH. The committee considers this system, which uses continuously stirred chemical reactors, to be a well-established technology for hydrolysis of chemical agent.

Treatment of Hydrolysates and Dunnage by Supercritical Water Oxidation (Steps 11 and 15)

Step 11 of the GATS process is treatment of the agent hydrolysate from the continuously stirred reactors by SCWO. **Step 15** is treatment by SCWO of the energetics hydrolysate mixed with hydropulped dunnage. The four SCWO units, two each for secondary treatment of each waste stream, are evaluated collectively.

Related SCWO Testing

Concurrent with the GATS EDS tests, the committee was also aware of other testing being conducted on SCWO reactor systems, including the following:

- testing of other SCWO reactor configurations, such as the transpiring-wall reactor (Crooker et al., 2000; Elliot et al., 2000; Griffith, 2000)
- the EST on a SCWO reactor one-tenth the size of the one anticipated for treating VX hydrolysate at the Newport Chemical Agent Storage Facility sponsored by the Project Manager for Alternative Technologies and Approaches, a part of the CSDP (NRC, 2001)

The committee also examined earlier laboratory-scale tests of materials of construction to address corrosion and fabrication problems encountered in applying SCWO technology to chemical demilitarization (General Atomics, 1997; SWEC, 1996). The SCWO reactor environment is highly oxidative, involves both high- and low-pH conditions, is highly turbulent in the reaction zone, and includes suspended solids. The 1997 General Atomics report describes corrosion studies on test coupons in a tubular reactor operated at 4,000 psig and 350°C (660°F) (subcritical), 450°C (842°F), and 550°C (1,022°F) with feed surrogates for mustard agent HD, neutralized HD hydrolysate, and HD hydrolysate with excess NaOH (General Atomics, 1997). Surrogates for HD and neutralized HD produced acidic test conditions; the HD hydrolysate with excess NaOH produced alkaline test conditions.

Materials classes that were tested included ceramics, nickel-based and cobalt-based alloys, refractory metals and alloys, reactive metals and alloys, noble metals and alloys, and high-temperature polymers, a total of 26 materials. Test periods varied between 37.5 and 47.5 hours. None of the materials was found to be suitable for all test conditions, and most exhibited moderate (equivalent to between 10 and 200 mil per year) to severe (>200 mil per year) corrosion. Titanium and titanium alloys (Nb/Ti and Ti-21S) exhibited the best performance, showing only slight corrosion in the presence of excess sodium hydroxide. Under acidic conditions, titanium showed increased rates of corrosion, apparently from attack by sulfuric acid and hydrochloric acid. Both localized pitting and wall thinning were observed.

Titanium corrosion also was observed during oxidation of cutting oil that contained sulfur and chlorine. For these tests, the SCWO system was operated at 600°C (1,112°F) and 3,400 psig (SWEC, 1996). Severe crevice and pitting corrosion were noted. In this study, acidic and reducing conditions (indicated by the presence of hydrogen sulfide) were present locally within the reactor. The operating period was approximately 50 hours.

GATS EDS SCWO Reactor Testing

The original General Atomics proposal for the GATS EDS included laboratory studies of materials of construction to evaluate how well different materials could withstand the SCWO environment. However, this part of the proposal was not accepted by the PMACWA, and materials of construction studies were not included in the EDS (M. Spritzer and G. Hong, General Atomics representatives, personal communication, December 13, 2000). To deal with the corrosive environment, the GATS SCWO reactor includes a replaceable inner liner. The EDS testing was done using liners made of platinum and two grades of titanium. Corrosion rates under various operating conditions were assessed.

The EDS testing was begun with a platinum reactor liner. When HD hydrolysate was processed, the liner showed significant corrosion. Corrosion rates as high as 10 mil per day were observed, making the replacement interval and cost of the 30-mil liners impractical. Thicker liners of platinum were considered too expensive. The high corrosion rate was attributed to attack by the chloride ions derived from the HD hydrolysate and by the formation of other platinum ions (M. Spritzer and G. Hong, General Atomics representatives, personal communication, December 13, 2000).

General Atomics subsequently tested a 110-mil-thick titanium liner and plans to test a 375-mil-thick liner for the full-scale reactor. Titanium is also subject to corrosion, but its lower cost makes a much thicker liner practical. General Atomics now plans to replace or invert this liner at 132-hr intervals, depending on how rapidly it corrodes (General Atomics, 2000c, 2000f, 2000g; Hong, 2001; M. Spritzer and G. Hong, General Atomics representatives, personal communication, December 13, 2000).

The corrosion occurs almost exclusively within five diameter lengths at the top of the SCWO reactor, in the vicinity of the injector nozzles. Very little corrosion is typically observed at the bottom. As a result, as the liner corrodes to the point where the top is at risk of failing, EDS tests demonstrated it could be inverted and used for approximately the same length of time again (Hong, 2001).

The corrosion rate of titanium in this service was fairly well characterized during the EDS testing and it is, indeed, quite high. Thus, General Atomics has come up with a scheme to replace a titanium liner at frequent enough intervals to allow for this corrosion rate. The committee points out that this is a mode of operation involving very high maintenance, which may take even longer on larger scale SCWO systems (see the discussion on "Maintenance Issues" later in the chapter). However, the committee cannot say that it is infeasible, since the technology provider has actually demonstrated it.

However, switching from platinum to titanium creates a new problem, because titanium tends to combust in pure oxygen. Consequently, the process design had to be modified to use air or synthetic air as the oxidizing medium. For synthetic air, pure oxygen is blended with pure nitrogen. If the mixing system fails to maintain the proper ratio, the potential for oxygen combustion remains. The lower oxygen and water activity resulting from the dilution with nitrogen, or the use of ordinary air, affects both the reaction rate and salt transportability. The lower reaction rate requires that throughput be reduced or that a larger reactor be used. The rate of nitrogen oxide formation from the use of nitrogen is expected to be too slow at the SCWO reactor operating temperature (616°C; 1,140°F) to be of concern. EDS test results on nitrogen oxide levels were not available when this report was prepared.

In addition to finding materials of construction that can withstand the corrosive SCWO environment, salt transport is essential to avoid plugging caused by salt deposition. Salts have a much lower solubility in supercritical water than in water at lower temperatures. The GATS SCWO system uses a proprietary additive to improve the mobility of the precipitated salt. In addition, the operating schedule calls for a rinse of the system with cooler, slightly subcritical water for 2 hours to dissolve salt deposits after every 22 hours of operation (Hong, 2001). This procedure appears to have been effective, and no problems with clogging or unacceptable pressure spikes were observed during the 500-hr test.

The EDS tests identified Hastelloy C-276 as a suitable material of construction for the following critical parts of the SCWO system:

- Performance was acceptable when the Hastelloy was used for the top insert of the reactor. The moderate rate of corrosion could be readily accommodated in this non-pressure-bearing part. The exposed area was small enough that effluent quality was not severely degraded by metals contamination.
- Good performance was exhibited in the quench zone of the reactor and when used for the removable sleeve for supporting the liner. The low rate of corrosion was consistent with long component lifetime.
- Excellent performance was shown in the region downstream of the reactor. No failures or indications of significant corrosion have been observed to date.

Processing and Treatment of Dunnage and Energetics Hydrolysate (Steps 12 to 16)

Step 12 of the GATS process is the separation, shredding, and grinding of the dunnage and other nonprocess waste until all solid material is reduced to granules <1 mm in size. These powdered solids are then mixed with energetics hydrolysate and other liquid wastes in a hydropulper to create a slurry that is then oxidized in a SCWO reactor.

It is critical that the feed to the hydropulper be reduced to a fine powder to avoid large particles in the slurry feed stream to the SCWO reactor. This may be difficult to accomplish for DPE suit material, because plastic tends to melt and stick to the shredder. Micronization of DPE suit material was successful during the EDS testing; however, several cycles of cryogenic cooling and micronizing were required to produce the desired particle sizes. This technology will require additional engineering design testing and validation before it can be used for the disposal facility at Pueblo.

Step 13 of the GATS process is the thermal treatment of metal parts from dunnage and DPE suits to a 5X decontamination level in an HDC. The committee considers this technology to be reasonably well established.

Step 14 of the GATS process is the slurrying of the size-reduced dunnage and nonprocess waste with energetics hydrolysate in preparation for feeding to the SCWO reactor. As discussed above, the size-reduction and slurrying equipment in the GATS design is commercially available and commonly used in many types of processing. With the possible exception of the DPE suits, the material can be shredded to the desired size and then slurried.

Step 15 is the treatment in the SCWO reactors of the micronized dunnage (with metal removed) slurried with energetics hydrolysate. The committee has concerns about how well the SCWO system can process slurries of organic materials, especially if they contain small quantities of tramp metal. The demonstration testing indicated very high metal-removal efficiencies from the shredded pallets. However, that high removal efficiency is unlikely to be maintained during actual operation. No SCWO testing was conducted using energetic hydrolysate slurried with shredded and micronized dunnage. This step remains to be demonstrated.

Step 16 of the GATS process is concentration of the brine from the SCWO reactors to reclaim the water and generate a solid salt cake for off-site disposal. Although the EDP did not include specific design parameters for this unit operation, it is existing technology, and, assuming that the SCWO reactor produces an effluent with the very low organic content called for in design specifications, appropriate concentration and crystallization equipment is commercially available. The committee notes that other chemical weapons demilitarization facilities have eliminated this processing step and suggests that a similar change be evaluated for the Pueblo facility.

ASSESSMENT OF INTEGRATION ISSUES

Component Integration

Destruction of the Pueblo stockpile to comply with provisions of the CWC treaty will require that the availability and throughput of each processing step, along with redundant process trains and sufficient buffer storage capacity between individual processing steps as necessary, achieve the destruction rate specified in the design package. General Atomics has not yet conducted a detailed throughput analysis that takes into account intermediate storage capacity. Such an analysis could assist in verifying that planned throughput rates can be achieved. Training is also important because the effectiveness of plant operating and maintenance personnel also contributes to process availability.

General Atomics has designed the GATS process and sized the equipment to process the Pueblo stockpile in 29 months (General Atomics, 2000a). The output rate from the reverse-assembly PMD system determines the size and number of units for all downstream process equipment. The GATS PMD equipment is similar to the equipment used in the baseline incineration system, and the operating experience from baseline facilities led General Atomics to conclude that a throughput rate of 50 lb/hr per machine would be attainable. General Atomics has determined that the long-term average capacity for the GATS design for Pueblo (actual throughput per year/maximum theoretical throughput per year) is 38 percent. To achieve this average capacity, two PMD machines are required to handle the Pueblo stockpile. The size and number of the rest of the General Atomics process equipment are planned to match the throughput of the PMD operation. For example, two SCWO reactors are used to treat the downstream agent hydrolysate, and two more SCWO units are used to treat the micronized dunnage and energetics hydrolysate waste stream.

Integrating the individual processing steps requires effective process monitoring and control to ensure that appropriate materials are fed to each processing step and that all materials discharged from the plant meet all safety and environmental specifications. Monitoring and control of the integrated facility using the GATS process will be based primarily on the strategies and means used in the baseline system. The overall monitoring and control system consists of the basic process control system (BPCS), the emergency shutdown system (ESS), and PLCs for individual equipment units. The BPCS comprises microprocessor-based controllers for monitoring and control. The ESS is a dedicated safety system of PLCs or microprocessor-based controllers that provide protective logic and shutdown capability. The means of controlling machines throughout a GATS facility are similar to those used for the baseline system machines (i.e., sequence-enabled functions with position switches).

Most of the monitoring instruments specified in the GATS design package are simple and reliable, having been used extensively in the chemical industry for many years. Control valves and monitors for temperature and pressure, as well as the distributed control systems and PLCs, have also been widely used in industry.

Process Operability

The operability of the SCWO reactors remains a significant issue. The reactors' operating conditions are set to balance competing conditions for minimizing plugging by salts and minimizing liner corrosion. That is, the conditions that result in good salt transport (and hence minimal plugging) are also the conditions at which corrosion is at a maximum. Conversely, operating conditions at which corrosion is at a minimum are conducive to the precipitation of salts that can cause plugging. General Atomics has approached this problem by (1) using a proprietary additive to improve salt transport and (2) designing the SCWO reactors with a slip-in sacrificial liner that would be replaced at regularly scheduled intervals. This combination, along with careful monitoring and control of temperature, pressure drop across the liner, additive feed rates, and other operating conditions, reduces the severity of the salt plugging and corrosion problems, but not sufficiently. The committee believes the SCWO system is still very difficult to operate, especially at full scale (see also the section on Maintenance Issues below).

Monitoring and Control Strategy

As discussed in previous NRC reports, except for the monitoring of corrosion and salt plugging discussed above, the GATS process does not require any unusual monitoring or control systems (NRC, 1999, 2000). The process control strategies consist of straightforward monitoring of pressure, flow rate, and temperature by well-established methods and equipment.

General Atomics believes that monitoring the turbidity resulting from titanium dioxide suspended in the reactor effluent will effectively monitor corrosion rates. Monitoring the turbidity of the effluent gives a good indication of the instantaneous corrosion rate, which can be used to ensure that operating conditions remain within the desired range. However, the decision to shut the process down for liner replacement would be facilitated if the extent of corrosion could also be monitored. This could be done by adapting one of the various probe designs available commercially or previously developed in other SCWO studies (Macdonald and Kriksunov, 2001). Another simple method would be measurement of the electrical conductivity of the thermocouple well.

Maintenance Issues

The EDS testing clearly showed that successful operation of the SCWO system requires an aggressive, proactive maintenance program to replace (1) the thermocouple well after approximately 60 hours of operation and (2) the titanium liner after approximately 130 hours of operation.

Replacement of internal components of the SCWO reactor is a time-consuming, elaborate procedure that involves cooling the system, flushing with clean water, manually removing the pressure head from the reactor, manually removing the liner, inverting the liner or replacing it with a fresh liner, reassembling the reactor, and restarting the system. General Atomics has performed this procedure more than 100 times on reactors with test-sized (3- to 4-inch ID) liners during the EDS and other SCWO test programs (Hong, 2001). During the EDS tests, the shutdown/start-up procedure required an average of 7 hours (3 to 11 hours) to complete.

However, maintenance has been performed only on comparatively small test-size SCWO reactors. The SCWO reactors proposed for the Pueblo facility are approximately 18 inches in diameter. The head at the top of an 18-inch reactor will not only be larger, but it will also have to be considerably thicker to withstand more than 20 times the internal pressure. Thus, the head on the full-scale reactor will be heavier and bulkier and will have more and larger bolts to be removed and replaced than the small SCWO test reactors. It will also have a larger sealing surface, which will have to be set and pressure tested. In the committee's experience, proper pressure testing of equipment is very time consuming. Consequently, the time and effort required to change the liner are likely to be much greater than for the EDS and Demonstration I tests. In the committee's opinion, maintenance approximately every 60 to 130 hours of operation will place a very heavy demand on operating and maintenance personnel.

Process Safety

The ACW I Committee concluded in its original report that there were "no unusual or intractable process safety problems" associated with the GATS process (NRC, 1999). However, in a subsequent evaluation of the Demonstration I test results, some aspects of process safety were identified (NRC, 2000). General Atomics also acknowledged the following safety design requirements in its report on the Demonstration I test results (General Atomics, 1999b):

- design modifications to incorporate equipment for removing aluminum hydroxide generated by the caustic dissolution of weapons to minimize aluminum-caused salt plugging and associated maintenance[4]
- control of volatile organic vapors generated in the ERH to prevent the accumulation of explosive mixtures in the ERH off-gas system and minimization of the maintenance requirements for removing condensed organics from fugitive emissions entering the ERH explosion containment cubicle
- safety features to preclude dust explosions in the DSHS

All of these concerns have been addressed in the EDS design package to the extent possible at this design stage (General Atomics, 2000a).

During the information-gathering phase of this report, the committee learned of an occurrence on December 2, 2000, involving backflow of fuel into the liquid oxygen feed line for the General Atomics SCWO reactor being operated in Corpus Christi, Texas, in conjunction with the EST for treatment of VX hydrolysate at the Newport site (PMACWA, 2000). This backflow resulted in overpressure (possibly in excess of 5,000 psig) and permanent expansion of part of the liquid oxygen feed line. The overpressure is believed to have resulted from the oxidation of fuel in the oxygen line. An earlier occurrence (July 14, 2000) involving the oxygen feed line resulted in a release of oxygen through the relief valve, whose metal components melted and started a grass fire (Bernard Bindle, safety engineer, PMCD, personal communication, March 26, 2001). The fire was attributed to removal of the high-pressure and high-high-pressure shutdowns from the pump circuit control and the use of stainless steel rather than monel for the pressure relief valve. Based on these incidents, the committee inferred that the General Atomics SCWO system would continue to be vulnerable to fires if pure oxygen and nitrogen were used to produce synthetic air for the SCWO reactors. The committee also noted that synthetic air could cause oxygen combustion of the titanium liner if the nitrogen-oxygen mixture was not controlled. Design modifications will be necessary to prevent similar

---

[4]For Pueblo, aluminum removal requirements are minimal because the only aluminum in the munitions is expected to be in fuze assemblies.

incidents in the future. Use of compressed air would completely eliminate this hazard.

As a part of the EDS design package, General Atomics prepared a preliminary hazards analysis (PHA) in accordance with Mil-Std-882C (DOD, 1993). The committee received a draft version of this package, and some committee members attended a design-review meeting that included a review of the draft PHA (General Atomics, 2000d, 2000h). The committee concluded that the PHA appears to be complete for the current stage of design and that key safety concerns have been identified for resolution in subsequent stages of design.

The PHA was based partly on interactive reviews by the engineers responsible for the EDS design package. These reviews were subsequently used to identify engineering measures for reducing risk. The committee believes that the use of procedural training and administrative solutions (e.g., checklists) for reducing risks should be minimized at this stage of design in favor of engineering design changes. The selection of hazard scenario frequencies in the completed PHA should be based on the assumption that procedures, training, and administrative controls established by the PMCD in charge of the CSDP will be used.

The PHA appears to have been conducted in a satisfactory manner at this stage of design, although it will have to be updated as the GATS design progresses, especially for parts of the design that are not yet fully developed, such as final designs of components supplied by the technology provider (ERH and PRH, for instance) and the selection of a method of supplying oxygen to the SCWO reactors that minimizes the risk of fire under upset conditions.

Worker Health and Safety

The conclusions regarding worker health and safety in the ACW I Committee's original and supplemental reports are still valid (NRC, 1999, 2000). The primary hazardous materials used during the destruction of agent and energetics are sodium hydroxide, liquid and gaseous oxygen, liquid and gaseous nitrogen, and methane (natural gas) for boiler fuel. Sodium hydroxide will be delivered in concentrated liquid form (50 weight percent) and diluted with water to produce lower strength solutions as required. This strongly corrosive caustic is handled safely in similar quantities and concentrations throughout the chemical industry and should not be unusually hazardous in the GATS process. Liquid and gaseous oxygen, liquid and gaseous nitrogen, and methane are also handled routinely and safely in many industries and do not represent an unusual hazard to workers.

As the design becomes more detailed, the PHAs will have to emphasize the safety of maintenance workers. The design should be configured to minimize the number of maintenance activities in contaminated areas to reduce worker risk and to ensure that worker access in DPE suits can be accomplished easily and safely. This precaution could also reduce the waste streams from used DPE suits and decontamination solution used during maintenance activities.

Public Safety

Accidental releases of agent or other regulated substances to the atmosphere or the groundwater system are extremely unlikely. Caustic scrubbing and activated carbon filtration are used on all gaseous process streams. Based on experience with the baseline facilities, these measures should provide a reasonable level of safety. Hold-test-release systems are not provided for gaseous effluents, but the scrubbing and filtration scheme, combined with the standard automatic continuous air monitoring system (ACAMS) monitors used at baseline facilities, should provide adequate protection against all gaseous process effluents. The facility HVAC design is similar to the design at existing baseline facilities, where air flows from clean areas to potentially contaminated areas and then through high-efficiency particulate air (HEPA) and activated carbon filters before release to the atmosphere.

The primary risk of a release of agent or other regulated substances is the explosion of a munition or the rupture of a pipe or vessel, but the likelihood of such an event for the full-scale facility should be extremely small. This conclusion is based on the committee's review of the General Atomics EDP for Pueblo and the understanding that the PMCD will require a comprehensive quantitative risk assessment (QRA) for the final facility design to ensure acceptable levels of risk. A QRA, which is much more detailed than the PHA performed at this stage of design, is a risk management tool during actual operation of the facility and serves as a basis for evaluating proposed changes in design or operation in accordance with CSDP risk management policies and procedures (PMCD, 1996, 1997).

QRAs are typically developed in parallel with the completion of facility design and construction. However, because design-based solutions to high-risk hazards can be implemented more easily during the design stage than during the construction stage, the committee recommends that the QRA process be implemented as early as possible. When a QRA done later reveals risks that must be addressed, the tendency is to rely on procedural/administrative solutions, which often complicate operations and are less effective than design modifications.

Human Health and the Environment

The environmental impact of the proposed GATS process appears to be minimal. All handling and processing of agent will be conducted indoors in sealed rooms that are vented through HEPA and carbon filters. Liquid and solid waste streams will be relatively small and manageable and will be subjected to hold-test-release procedures.

Effluent Characterization

The liquid effluent, which consists of water from the evaporator/crystallizer used to produce the solid filter cake produced by the brine-recovery operation, should not pose a significant hazard to human health or to the environment. Much of the recovered water is recycled for use in the process. The solid waste from the process, consisting of dried filter cake, is likely to require stabilization prior to disposal in a hazardous-waste landfill.

The EDS tests indicate that the hazardous constituents of gaseous effluents are present only in very low concentrations, especially in the gaseous effluent from the SCWO reactor liquid after pressure letdown. If scrubbers and carbon filters are used properly, these discharges should meet regulatory standards. However, this must be confirmed through comprehensive testing and characterization of the trace constituents of gaseous effluents.

Completeness of Effluent Characterization

The liquid and solid effluents are well characterized, but only major constituents of the gaseous effluents have been characterized. As the ACW I Committee noted in its original and supplemental reports, the gaseous process emissions will have to be characterized for the health and environmental risk assessments required by EPA guidelines (NRC, 1999, 2000). Measurements to date are not adequate for the committee to evaluate the environmental impact of the process. Standard EPA methods of analyzing samples of gaseous effluents generally produce full scans that can indicate the quantities of a large number of compounds of environmental concern.[5] These results, along with the results of analyses of emissions of metals (including chromium VI), can be used to assess the environmental impact of a facility through accepted risk-assessment methods (EPA, 1998).

Effluent Management Strategy

The proposed strategy appears to be reasonable and should protect public health and the environment.

Off-site Disposal Options

Dunnage. Experience at JACADS and Tooele has shown that only a tiny fraction of dunnage is contaminated with agent. Uncontaminated dunnage from these two stockpile locations is being disposed of off site by commercial waste management facilities (McCloskey, 2000; U.S. Army, 1998). Off-site management of uncontaminated dunnage is also planned for both the Newport and Aberdeen sites. Off-site management of dunnage from Pueblo would greatly reduce the on-site processing requirements and greatly simplify process integration by eliminating the need for size reduction and SCWO treatment of the dunnage.

Brines. Brines produced from air pollution control processes at Tooele are currently being shipped off site for disposal by commercial waste management facilities. The Army also plans to ship the effluent from SCWO (after concentration by evaporation) at Newport off site for disposal. This material has been delisted as a hazardous waste by the state of Indiana, and the Army has identified 16 commercial facilities that can accept the brine (Wojciechowski, 2000). Off-site management of SCWO effluent after evaporation (to recover water) would eliminate the need for a crystallizer and simplify process integration at Pueblo.

Environmental Compliance and Permitting

The combination of technologies in the General Atomics technology package is not expected to lead to problems with environmental compliance or permitting. All process waste streams except the SCWO off-gas will be evaluated prior to release to confirm that they are either free of regulated substances or that these substances are at acceptable concentrations. The SCWO off-gas will be scrubbed, monitored by ACAMS, and passed through activated carbon filters.

ASSESSMENT OF OVERARCHING TECHNICAL ISSUES

Overall Engineering Design Package

Although the EDS test results with the PRH, the ERH, and their HDCs appear to warrant proceeding with additional developmental testing of these GATS components, the corrosion of the SCWO reactor liner raises questions about whether the GATS process could destroy the munitions stored at Pueblo within a reasonable period of time. Even though tests of the dunnage-shredding and micronizing system were successful, this system would be superfluous if the SCWO system cannot treat the resulting dunnage slurry. The SCWO system, a major part of the process, still has significant problems, especially the high maintenance associated with corrosion of the reactor liner and thermocouple well.

Steps Required Before Implementation

The initial evaluation of the GATS process by the ACW I Committee identified the following steps required for implementation (NRC, 1999). These steps are reevaluated below.

> Conduct tests of the cryofracture process to ascertain if it provides better access to the agent cavity in projectiles and mortars than the baseline disassembly process.

---

[5]EPA analyses are done with 8000 Series methods, especially those using gas chromatography/mass spectrometry scans—for example, Methods 8260B, "VOCs by GC/MS," and 8270C, "Semi-VOCs by GC/MS."

The cryofracture process was not tested in the EDS program. Therefore, concerns about operating hazards associated with the possible accumulation of agent in the cryobath have not been addressed. In addition, the advantages of cryofracture over the baseline system disassembly process have not been demonstrated.

> Sample and analyze air emissions from the demonstration system. The air emissions will have to be measured to a level of detail and accuracy that can be used for HRAS (health risk assessments) and environmental risk assessments required by EPA.

Subsequent testing resulting in extensive analyses of air emissions revealed no obvious problems. However, effluents from all major feed streams in the GATS process for Pueblo have not been characterized (e.g., effluent from the SCWO reactors that process the energetics hydrolysate-dunnage slurry) because EDS testing did not include processing of those streams. Moreover, final determinations of safety and environmental acceptability can only be made through a formal risk assessment process.

> Verify that energetic materials encased in metal (e.g., rocket or other munitions fragments) will be hydrolyzed.

The EDS testing appears to demonstrate that energetic materials encased in metal can be hydrolyzed successfully, although some questions remain regarding the absolute completeness of the hydrolysis. During the Demonstration I testing, fuzes were observed to have popped on the HDC, indicating that they may not have been completely hydrolyzed. Also, the ERH tests were conducted on rocket segments even though there are no rockets in the stockpile at Pueblo.

> Ascertain how well the SCWO process can handle high-solids materials (shredded dunnage).

No SCWO testing was conducted using energetic hydrolysate slurried with shredded and micronized dunnage. Consequently, this step remains to be demonstrated.

> Determine erosion and corrosion behavior of the components of the SCWO system.

The EDS testing and other testing cited in this report provided design information on the erosion and corrosion behavior of SCWO components. The data, which appear to be excellent, confirm high rates of corrosion. As discussed elsewhere in this report, the GATS SCWO system, as currently designed, would present a significant maintenance burden for operators and maintenance personnel.

Previous Findings and Recommendations

In this section, the findings and recommendations regarding the GATS process from the two ACW I Committee reports are reviewed to determine if they are still valid or if they have been addressed by the EDS tests or by information from the evolving GATS design (NRC, 1999, 2000).

Review of Findings from the 1999 Initial ACW I Committee Report (NRC, 1999)

Finding GA-1. Cryofracture appears to be an effective method for accessing the agent in projectiles and mortars and might provide an improvement over baseline disassembly in accessing gelled or crystallized agent. This remains to be demonstrated.

Cryofracture as a means of accessing the agent cavity has still not been tested. The ACW I Committee's original finding remains unchanged (see new Finding [Pueblo] GA-8 and Recommendation [Pueblo] GA-4).

Finding GA-2. Hydrolysis of energetics at the scales proposed by the technology provider is a relatively new operation. Chemically, it is possible to hydrolyze all of the energetic materials; however, the rate of hydrolysis is limited by the surface area and, therefore, depends on particle size. (Smaller particles are more desirable because they have a higher surface-to-volume ratio.) The proposed method of removing and hydrolyzing the energetics appears to be reasonable, but further testing is required to determine the hydrolysis rates and to confirm that throughput rates can be achieved.

This finding is still valid. The ERH successfully hydrolyzed fuzes and bursters (Demonstration I), as well as 4-inch to 8-inch rocket segments of propellant (EDS). The hydrolysis time for rocket segments is consistent with overall process feed rates. Further process development is ongoing.

Finding GA-3. The rotary hydrolyzer appears to be a mature reactor configuration that is well suited for this application.

The committee's evaluation of the detailed design for the ERH raised a concern about basing the design of a full-scale flow ERH strictly on data collected from ERH testing conducted in a batch reactor. This finding is superseded by the new finding (Pueblo) GA-2.

Finding GA-4. Shredding of dunnage and injection of the slurry directly into a SCWO system is a new and unproven process. General Atomics claims to have developed a proprietary pump capable of pumping the slurry at high pressures, but it has not been tested under the intense solids loading anticipated. Furthermore, the injection of large amounts of solid material, including shredded wood, cut-up nails, and complex organic materials, such as pentachlorophenol and other wood preservatives, into the SCWO system has not been demonstrated. Considering the difficulty SCWO reactors have encountered with deposition of solids when liquids are treated, the committee believes that this application of SCWO may encounter significant difficulties. (At the time of this writing, processing of solids with SCWO was being performed as part of the ACWA demonstrations.)

The EDS testing has demonstrated the dunnage-shredding system with some success. However, no tests have been conducted for SCWO treatment of shredded and slurried dunnage. As a result, this finding remains unchanged.

Finding GA-5. All of the findings in the NRC report, *Using Supercritical Water Oxidation to Treat Hydrolysate from VX Neutralization*, apply to the General Atomics system.

A subsequent NRC report indicated that the findings presented in the 1998 NRC report *Using Supercritical Water Oxidation to Treat Hydrolysate from VX Neutralization* are still applicable to the SCWO system planned for Newport (NRC, 2000). For Pueblo, considerable additional data have been accumulated from Demonstration I and EDS testing, including the data resulting from a planned change in materials of construction from platinum to titanium. Nevertheless, concerns about the SCWO system for Pueblo remain.

Finding GA-6. The crystallization and evaporation operations have not been tested for this application. Although these are conventional technologies and are expected to work effectively, testing will be necessary.

No testing of crystallization and evaporation has been performed. This finding is unchanged.

Finding GA-7. No hold-test-release facilities are provided for gases from the hydrolysis reactors or the SCWO reactors. These gases will be scrubbed using activated carbon prior to release.

This finding is unchanged; however, the committee considers this requirement to be unnecessary for safety or protection of the environment.

Review of Findings and Recommendations from the 2000 ACW I Committee Supplemental Report

Finding (Demo I) GA-1. Testing on the hydrolysis of energetic materials contaminated with agent will be necessary before a full-scale system is built and operated.

This finding is not being addressed by the ACWA EDS program.

Finding (Demo I) GA-2. Testing will be required to verify that the larger-diameter supercritical water oxidation (SCWO) reactor feed nozzles will be capable of accepting the dunnage material as shredded (i.e., without additional classification and segregation) and that the reactor will perform reliably under these conditions.

This finding is still valid. During EDS testing, micronizing and slurrying of the dunnage was successful, but no SCWO testing with this slurried material was performed.

Recommendation (Demo I) GA-1. Operation of the size-reduction and slurrying system, and long-term operation of the supercritical water oxidation (SCWO) reactor with slurry, should be conducted before proceeding with a full-scale system.

This recommendation has been partially fulfilled. The size-reduction system appears to have performed well in both Demonstration I and EDS testing; however, SCWO treatment of slurried dunnage has not been tested to date.

Recommendation (Demo I) GA-2. Before construction of a full-scale supercritical water oxidation (SCWO) system, additional evaluations of construction materials and fabrication techniques will be necessary because corrosion and plugging prevent continuous operation with the present design. If the new construction materials do not solve these problems, then alternative SCWO reactor designs should be investigated.

The EDS test results reinforce this recommendation. Furthermore, scale-up from the test vessel, with a diameter of approximately 4 inches, to a full-scale reactor, with a diameter of 18 inches, introduces additional uncertainties about how well the reactor liner will withstand corrosion.

Recommendation (Demo I) GA-3. To determine the operability of the supercritical water oxidation (SCWO) reactor and the reliability of the materials of construction, long-duration runs of a SCWO reactor should be conducted with slurry, with energetics hydrolysate, and with agent hydrolysate before full-scale implementation proceeds.

This recommendation is still valid.

Recommendation (Demo I) GA-4. The efficacy and safety of the additional step to remove aluminum hydroxide from the hydrolysate produced from rocket propellants should be evaluated prior to construction of a full-scale supercritical water oxidation (SCWO) system.

This recommendation is not pertinent to the Pueblo facility because no rockets are stored at the Pueblo site. EDS testing showed that phosphate precipitation did remove aluminum from the hydrolysate. However, the testing was of insufficient duration to ascertain whether the removal was adequate for the small quantities of aluminum in the munitions stored at Pueblo.

Recommendation (Demo I) GA-5. Decontamination of solid munitions materials by flushing and immersion should be demonstrated prior to full-scale implementation.

This recommendation is still valid. No testing of agent-contaminated munition material has been conducted to date.

Recommendation (Demo I) GA-6. The air emissions data from the demonstration tests should be used in a screening risk assessment. The results of the air effluent samples should be subject to (1) a human health risk assessment following the Human Health Risk Assessment Protocol (HHRAP) for Hazardous Waste Combustion Facilities from the Environmental Protection Agency (EPA) [EPA530-D-98-001(A,B,C)] and (2) an ecological risk assessment following a protocol that will be released by EPA in the very near future.

This recommendation is still valid. At the time this report was being prepared, no environmental health or ecological risk assessments for a disposal facility at Pueblo had been performed based on the available emissions data from the EDS testing (General Atomics, 2001).

New Findings and Recommendations

Finding (Pueblo) GA-1. The GATS projectile/mortar disassembly (PMD) machines should perform as well as they do in the baseline system. The modifications in the reverse assembly steps in the GATS process should not substantially affect the performance of the PMD machines.

Finding (Pueblo) GA-2. The GATS energetic rotary hydrolyzer (ERH) is a new technology that has been tested only in a batch-system mode. None of the testing to date has been performed on a full ERH operating system in a flow-system mode in conjunction with a heated discharge conveyor.

Finding (Pueblo) GA-3. The liquid nitrogen tank of the cryofracture system could be a site of accumulation for neat agent that may leak from munitions either because the munitions are already leakers or because thermal shock causes a leak. Although it is unlikely that very much agent would leak from any one munition, small amounts could potentially accumulate in the liquid nitrogen bath and create a hazard during cleaning and other maintenance activities.

Finding (Pueblo) GA-4. Additional testing and demonstration of the demilitarization protective ensemble (DPE) suit shredding and micronizing system will be necessary to establish the operability of the process at full scale. The EDS tests established the proof of concept but did not demonstrate the operability of a continuous shredding and micronizing system for DPE suit material.

Finding (Pueblo) GA-5. The GATS supercritical water oxidation (SCWO) system is subject to large, sudden pressure swings that could cause a backup of materials from the SCWO reactor into the feed lines. If pure oxygen is one of the feed streams, pressure upsets could create a safety problem. During engineering-scale testing of a General Atomics SCWO reactor configuration intended for use at the Newport site to treat VX hydrolysate, flammable material forced by pressure surges into an oxygen line caused a fire.

Finding (Pueblo) GA-6. Although the operation of the GATS supercritical water oxidation reactor was demonstrated over a 500-hr period during the engineering design study tests, the reactor required frequent shutdowns for inspection, maintenance, or replacement of corroded reactor components. This high maintenance requirement would seriously burden operating personnel in a full-scale operation.

Finding (Pueblo) GA-7. Corrosion remains a serious operating problem with the GATS supercritical water oxidation system. Failure to shut down in time to replace a perforated reactor liner could result in rapid corrosion of the high-pressure reactor shell.

Finding (Pueblo) GA-8. A detailed throughput analysis that takes into account intermediate storage capacity has not been carried out. A throughput analysis would verify if planned throughput rates can be achieved.

Recommendation (Pueblo) GA-1. Provisions should be made in the design of the liquid nitrogen cryobath for the remote cleanout of residues that may be contaminated with agent.

Recommendation (Pueblo) GA-2. A flow-system version of the energetics rotary hydrolyzer should be built and tested to assess materials flow characteristics prior to construction of the full-scale system. The tests should be performed on chunks of material similar in size to the fuzes, bursters, propellants, and mixtures of energetics and metals and other inert components.

Recommendation (Pueblo) GA-3. Methods for monitoring the extent of corrosion in the supercritical water oxidation reactor other than by measuring effluent turbidity should be investigated. Outputs from the monitors should identify corrosion that exceeds safe operating limits, enabling operators to take corrective action when necessary.

Recommendation (Pueblo) GA-4. Safeguards should be provided for the GATS supercritical water oxidation (SCWO) system against pressure surges causing material from the reactor to flow back into a feed pipe. The use of pure oxygen feed for the GATS SCWO system should be avoided if at all possible.

4

Parsons/Honeywell Technology Package

DESCRIPTION OF THE PROCESS

Introduction and Overview

The Parsons/Honeywell team includes Parsons Infrastructure and Technology Group, Inc.; Honeywell, Inc.; the Illinois Institute of Technology Research Institute (IITRI); and General Atomics, Inc. The team uses the acronym WHEAT (water hydrolysis of explosives and agent technology) to denote its technology package for the demilitarization of assembled chemical weapons. The process proposed for disposal of the stockpile at the Pueblo Chemical Depot includes seven basic operations:

- The Army's baseline disassembly process, with modifications, to separate agent, energetics, and metal parts.
- Batch hydrolysis of mustard agent (HD and HT).
- Batch hydrolysis of energetics.
- Biological processing, followed by evaporation/crystallization, to convert the hydrolysis products to liquids or solids acceptable for discharge to the environment or liquids acceptable for recycling. Biological treatment is done in the immobilized-cell bioreactor (ICB).
- Thermal treatment in the batch metal parts treater (batch MPT) or the rotary metal parts treater (rotary MPT) to decontaminate metal parts to 5X.
- Heat treatment in the continuous steam treater (CST) to decontaminate dunnage to 5X.
- Catalytic oxidation for cleansing process gas discharges from the plant and activated carbon filtration for some or all of the off-gas streams.

A block flow diagram for the Parsons-Honeywell technology package is presented in Figure 4-1, and a detailed description of the package follows.

Disassembly of Munitions and Removal of Agent and Energetics

The munitions to be processed from the Pueblo Chemical Depot stockpile are listed in Table 1-2. The Parsons/Honeywell process is designed with the expectation that on-site containers, such as those used to transport munitions from the Deseret Chemical Depot storage area in Utah to the adjacent disposal facility at Tooele, will not be used (Parsons, 2000b). Instead, modified ammunition vans (MAVs) are used to transport munitions from the storage area to the disposal facility. Moreover, the Parsons/Honeywell technology package does not include a container-handling building. Munitions are transported from the depot storage igloos directly to the on-site munitions storage building (MSB) and from there to the unpack area (UPA) in the munitions demilitarization building (MDB) (Parsons, 2000c). Munitions disassembly involves the following areas and systems:

- MSB (buffer storage area) and a UPA
- WHEAT projectile/mortar disassembly (WPMD) machine
- WHEAT multipurpose demilitarization machine (WMDM)
- energetics rotary deactivator (ERD)
- burster washout machine (BWM)
- energetics shredder
- projectile rotary washout machine (RWM)

The normal sequence of events for disassembly can be summarized as follows. Munitions are removed from their pallets or boxes in the UPA and, if they contain no propellant, are conveyed into the explosion containment room (ECR), where the WPMD machine begins disassembly. Bursters and burster tubes are removed and sent to the BWM, where energetics are removed and fed to the energetics shredder. The shredded energetics are then sent to the ener-

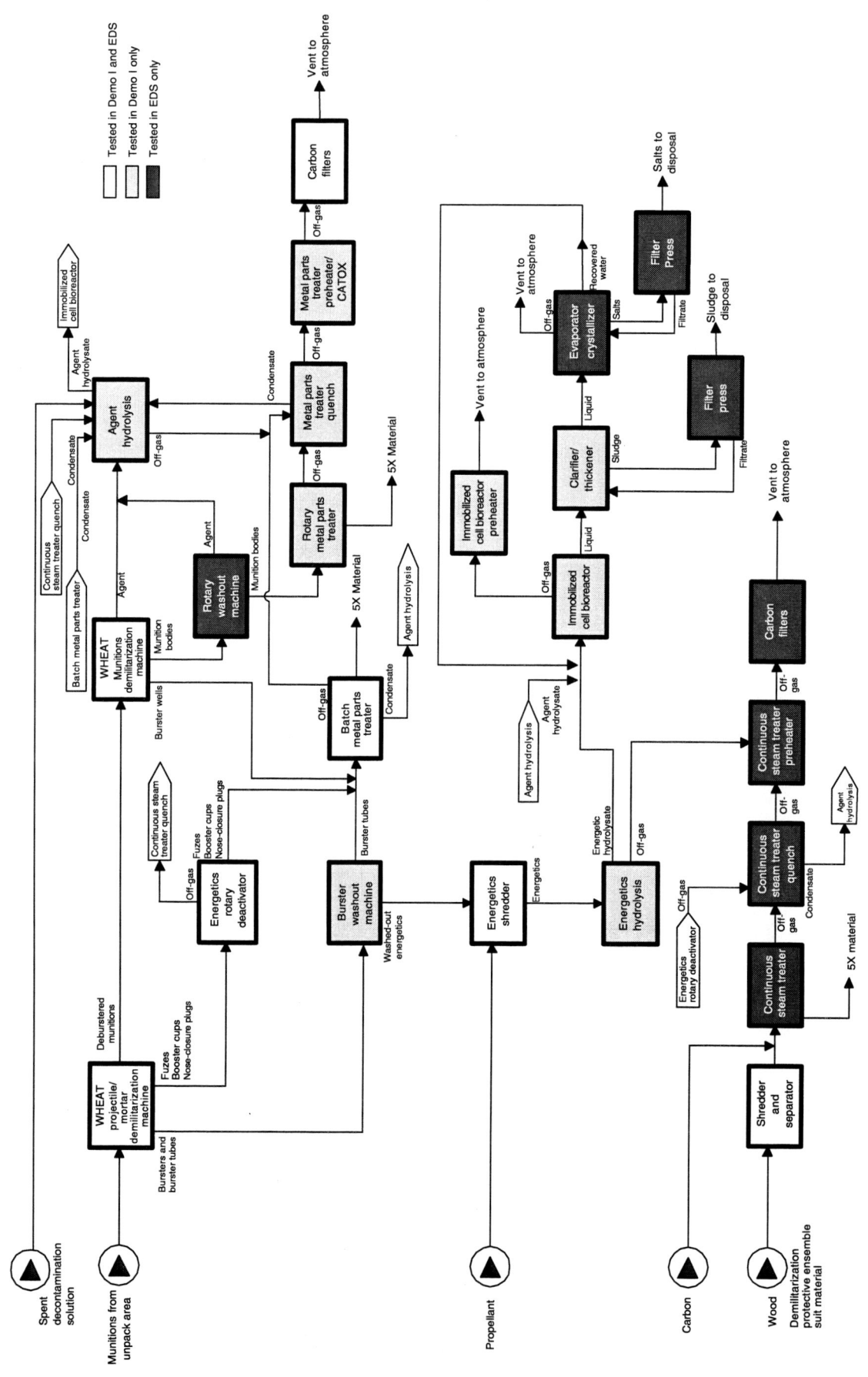

FIGURE 4-1 Parsons/Honeywell WHEAT block flow diagram. Source: Adapted from Parsons, 2000a.

getics hydrolysis system. Fuzes, burster cups, and nose-closure plugs are also removed in the WPMD and are fed to the ERD. The deburstered munitions from the WPMD are sent to the WMDM, where the burster wells are removed and the agent is drained to the agent storage tank. The burster wells are fed to the batch MPT for 5X treatment, and the munition bodies are sent to the projectile RWM. Any residual mustard is removed from the agent cavity by washing with high-pressure water jets in the RWM, and the washed bodies are sent to the rotary MPT. The spent wash solution from the projectile RWM is sent to the agent hydrolysis system. More complete descriptions of these operations follow.

Buffer Storage Area and Unpack Area

Because nighttime transport of munitions from depot storage to the plant site will not be allowed, 24 hours of buffer storage capacity for incoming munitions will be provided in the MSB. If thawing of the mustard agent in the incoming munitions is necessary, it will take place in the MSB (winter and summer design temperatures for the Pueblo facility are –20°F and 97°F [–29°C and 36°C], respectively). Munitions that have been identified as leaking chemical agent (leakers) will not be included in the normal feedstock retrieved from the storage igloos. Known leakers will be processed separately, as they are at baseline system facilities. However, during the energetics removal process, especially upon removal of the burster tube, a leak may be detected or may develop. In such cases, the munition is to be decontaminated, overpacked, and moved to a separate area for further processing. In the UPA, propellant-free munitions delivered from storage igloos are removed from their pallets.

Pallets of boxed munitions that contain propellant are also unpacked in the UPA (Parsons, 2000c). Munitions are then moved to the propellant removal room (PRR), where the munition inside its fiberglass container is placed in a glove box and monitored for agent. If agent is detected, the munition is overpacked and sent back to the depot for storage and will be demilitarized in a separate campaign. If agent is not detected, the munition is removed from the container and moved to one of four stations, where the propellant is removed; in the case of 4.2-inch mortar rounds, both the propellant and the ignition cartridge are removed. The propellant is temporarily stored in a propellant storage room and eventually returned to the depot for storage. The propellant-free munition is either returned to the depot for storage or returned to the UPA for disassembly in the ECR.

Projectile/Mortar Disassembly

Projectiles are disassembled in individual campaigns for each caliber of munition by the WPMD, an eight-position, rotating table machine with five main stations. The WPMD removes the nose-closure/lifting ring or the fuze (for 105-mm projectiles and mortars, respectively). Fuzes with booster cups are removed and punched by the WPMD to expose the explosive. The 4.2-inch mortar has a steel burster well attached to the fuze by a threaded connection. The WPMD unscrews and removes this assembly and then extracts the burster from the burster well. For all other projectiles, the burster in its metal casing is removed from the burster well by another WPMD station. The bursters are fed to the BWM; mortars and projectiles that have had their bursters removed are fed to the WMDMs in the munitions demilitarization machine (MDM) room. The nose-closure parts, fuzes, booster cups, and miscellaneous parts are fed to the ERD.

Energetics Rotary Deactivator

When the ERD receives the fuzes, booster cups, and miscellaneous parts from the WPMD, it heats them until they are deenergized (i.e., until they deflagrate or detonate). The shell of the ERD is maintained at 1,250°F (677°C) by electric-induction heating, and the parts are heated to 650°F (343°C) in about 5 minutes. One ERD is located in each of the two ECRs. The deenergized fuzes and booster cups then exit the ERD onto a conveyor that moves them to the batch MPT for 5X decontamination. Before exiting the ECR, washed burster tubes from the BWM are added to the same conveyor. In the parts transfer area, the conveyor material drops into a container, which is placed on the batch MPT conveyor by a pick-and-place machine. En route to the batch MPT, the containers stop to receive burster wells that have been removed at the WMDM.

During campaigns for 155-mm projectiles, the lifting lugs are fed to the ERD. At those times, the ERD is used only to move materials, and the induction heating coils are not activated. The vent gases from the ERD are sent to the MPT quench tower.

Burster Washout Machines

Bursters from the mortars and projectiles are fed into the BWMs by a pick-and-place machine and processed in the BWMs to wash out all explosives (Parsons, 2000c). There is one BWM in each ECR (total of two). The BWM has a rotary carousel with multiple receptacles. Bursters are aligned with a multinozzle water-jet washout probe on the BWM so that the jet cuts into the explosive charge axially from the open end. The water jet, which contains no abrasive, is injected at about 12,000 psi, although lower pressures of 2,000 to 3,000 psi are being considered. When the jet reaches the end of the burster tube, the water-jet probe is withdrawn. The washed burster tubes are discharged from the BWM one at a time onto a conveyor for transport to a container-loading station in the parts transfer area and then conveyed to the batch MPT for 5X decontamination.

The washout water from the BWM entrains the explosive particles and washes them out of the burster casing. The particles and water pass through a low-speed shredder that reduces all particles to <0.25 inch in diameter, which facilitates the transport of the resulting slurry to the hydrolysis reactors and lowers hydrolysis reaction residence time. The slurry is discharged from the shredder to a collection tank. From there, it is pumped to the hydrolysis reactors.

Multipurpose Demilitarization Machine

The agent-accessing process in the Parsons/Honeywell EDP is a new design that has a WMDM and an RWM (Parsons, 2000c). The WMDM functions much like the baseline system MDM but is designed to contain agent spillage, which sometimes occurred during disposal operations at JACADS when the projectile burster wells were pulled.

The Parsons/Honeywell munitions-processing scheme is also intended to solve the problem of draining partially solidified agent, which has been found in mustard-filled munitions. Draining of mustard-filled munitions at JACADS was problematic because of the presence of an unpredictable quantity of degradation products in the form of mustard sludge/solids. The first operation of the WMDM is removal of the burster well by a pull station that has a cylindrical containment/splash-guard attachment to contain spillage. The burster wells removed from the munitions bodies are placed in the energetics parts containers for processing in the batch MPT. The WMDM has a station for cutting through the casing wall in the eventuality of a failed pull operation. The baseline draining station has been replaced in the Parsons/Honeywell design with a tilt-and-drain station to remove agent that is liquid. The drained casings are then fed forward to the projectile RWM, where sludge and solids are removed. The drained agent is then collected and transferred to agent storage tanks in the toxic cubicle.

Projectile Rotary Washout Machine

The solid heel or sludge that remains inside the munitions casing is washed out in the projectile RWM using recirculated wash water through high-pressure water jets (Parsons, 2000c). The optimum temperature for the water-jet washout of munitions has not yet been determined, but the guideline is to keep the water temperature low so the bulk of the mustard materials inside the munitions are washed out with minimal hydrolysis of the mustard. Parsons expects the temperature to be between 16 and 43°C (60 and 110°F) (Parsons, 2000d). The washout solution is acidic to minimize agent hydrolysis.

The agent washout slurry is then allowed to settle. The supernatant water, which contains dissolved thiodiglycol and hydrochloric acid, is recycled to the water-jet probes. A wash-water purge stream is mixed with fresh make-up water in an approximate ratio of 1:3 (wash-water purge: fresh water) and fed to the hydrolyzers. The settled material, anticipated to be 90 percent mustard, is removed and sent to the holding tank for agent concentrate in the toxic cubicle.

The washed munitions, which are expected to have no more than 2 volume percent agent in the cavity, are delivered to a conveyor and moved through an airlock to the loading device of the rotary MPT. Each munitions-processing line has its own RWM.

Hydrolysis of Agent and Energetics

Hydrolysis of Agent

A flow diagram of agent neutralization in the Parsons/Honeywell WHEAT process is shown in Figure 4-2. Drained agent from the WMDM, agent concentrate from the projectile RWM, MPT condensate, CST condensate, and spent decontamination solution from the agent and energetics hydrolyzers are stored in three cubicles in the MDB. Up to 5,300 lb of agent is stored in a 500-gallon tank. A 1,300-gallon tank is also available but is intended for emergency use only. Agent concentrate (up to 4,680 lb of mustard as a nominal 90 weight percent mustard solution) is stored in another 500-gallon tank. These three tanks are in the toxic cubicle. The MPT and CST condensate streams, which are the purge streams from the MPT and CST quench towers, are stored in two more tanks, each with a 6,800-gallon storage capacity. Once these streams have been shown to be free of agent, they are sent to the agent hydrolysate tank. If agent is present, they are sent to the agent hydrolyzers. The tanks holding condensate provide about 24 hours of buffer storage. The spent decontamination solution is sent to the agent hydrolysis reactors.

The hydrolysis of agent is carried out in three identical agent-neutralization rooms (Parsons, 2000c). Each room contains two agent-hydrolysis reactors and one holding tank for spent decontamination solution. The agent hydrolyzers are used to destroy agent drained from the WMDM, agent concentrate from the projectile RWM, and any agent in the spent decontamination solution. They are also used to destroy agent in the MPT/CST condensate, if any has been detected.

Agent from the holding tank in the toxic cubicle (drained agent) is pumped, along with hot water, through a static-mixer eductor, which disperses the agent in the water. The dispersion is pumped to a well-stirred 2,520-gallon (1,525-gallon working capacity) reactor lined with polyvinylidene fluoride and jacketed with hot water. The reactor is partially filled with hot water at 90°C (194°F) (titanium reactors are being considered as alternatives to PVDF-lined reactors). The agent concentration in the reactor is approximately 4 percent. The agent feed rate is controlled to maintain an excess of water, which prevents the formation of sulfonium salts that would slow the hydrolysis and give rise to additional by-products. As the agent reacts with water, hydrochloric acid is produced, which lowers the pH to about 2.

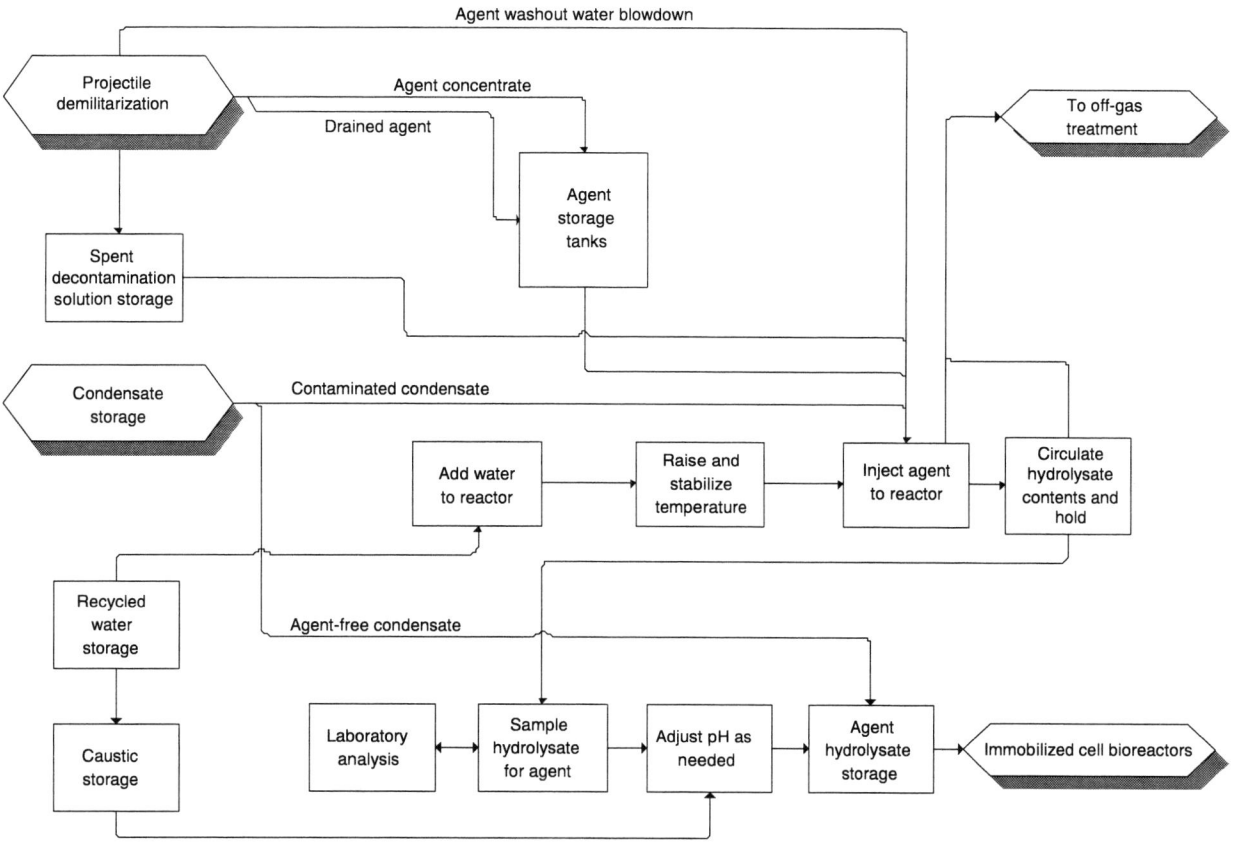

FIGURE 4-2 Agent hydrolysis process. Source: Adapted from Parsons, 2000a.

The reaction time is about 2.5 hours. Because the reaction is exothermic, heat is removed through a jacket and an external loop with a heat exchanger. Upon completion of the reaction, 18 percent sodium hydroxide is added to adjust the pH to between 10 and 12, preventing the reforming of agent. The hydrolysate is then transferred to the hydrolysate tank.

The reactor is maintained at 90°C (194°F) during hydrolysis and blanketed with nitrogen. The pressure is maintained at 3 psig, and the reactor is vented to the MPT off-gas treatment system except during agent filling, when the vent is closed.

The batch cycle for the hydrolysis of agent concentrate from the projectile RWMs is the same as for the drained agent. Spent decontamination solution is pumped from holding tanks to the hydrolyzers as required.

Hydrolysis of Energetics

Figure 4-3 is a flow diagram of energetics neutralization in the Parsons/Honeywell process. The feed to the energetics hydrolysis system is a slurry of burster energetics consisting of tetryl, tetrytol, or a mixture of TNT and tetryl, depending on the type of munitions being processed (Parsons 2000a, 2000c). The slurry particles are 0.125 inch in diameter or less. Propellants will be hydrolyzed during the facility close-out campaign (Parsons, 2000d). Three reactors, each with an operating capacity of 200 gallons, are installed, with two in operation and one as a spare. The reactors are constructed of 304 stainless steel and have jackets for heating and cooling. Vigorous agitation is provided in the reactor to keep the energetics particles suspended. An external pump-around loop may be used to provide additional agitation; a decision on this possible design feature is pending. The batch cycle starts with the addition of caustic and water to the reactor. The temperature is then raised to 90°C (194°F), and the energetics slurry is added, achieving a caustic-to-TNT or caustic-to-tetryl ratio of 3:1 and an energetics loading of 12 weight percent. An antifoam agent is also added. The batch cycle hydrolysis time is 8 hours. At the end of the reaction period, the batch is sampled and the pH adjusted to within a range appropriate for biotreatment (see next section) if required. The product is pumped to an energetics hydrolysate storage tank and then fed to the bioreactors. Off-gases from the reactor are sent to the quench tower for the rotary MPT.

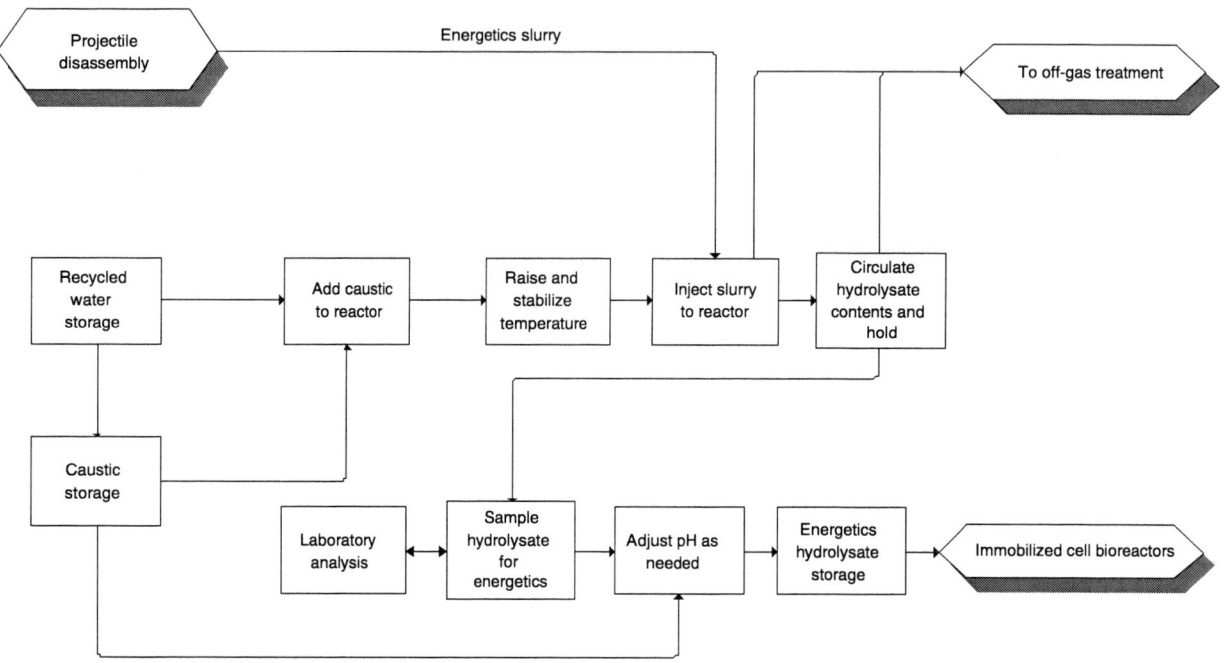

FIGURE 4-3 Energetics hydrolysis process. Source: Adapted from Parsons, 2000a.

Biological Treatment

In the Parsons/Honeywell biotreatment system (shown in Figure 4-4), the agent and energetics hydrolysates are combined and diluted with water, mixed with inorganic nutrients, and fed to the ICBs (immobilized-cell bioreactors), which contain aerobic microorganisms that will consume most of the organic content of the hydrolysates. The bioreactor system consists of 16 ICBs arranged in four modules, with four ICBs each. Each ICB has a capacity of 40,000 gallons of liquid and a residence time of 5 days. Each ICB is fed 1,600 standard cubic feet per minute (scfm) of aeration air from a 6,400-scfm air blower common to the four ICBs in a module. The allowable temperature range of the liquid feed is 24°C to 46°C (75°F to 115°F), with a pH of 6 to 8 maintained by adjusting with caustic. Parsons/Honeywell plans to explore using higher concentrations of agent and energetics in the feed to the bioreactors to reduce the number of bioreactors from 16 to 12.

Each ICB has three chambers, with air sparged into the bottom of each chamber. Agent and energetics hydrolysates are mixed with nutrients and process water and fed to the bottom of the first chamber. The air and liquid flow concurrently up through a packed bed in the ICB. The liquid then flows to the second chamber and then to the third. A microbial culture, specific to the organic constituents in the feed, is established in the ICB packed-bed media. This culture digests the organics, producing carbon dioxide and water, as well as other typical waste-treatment effluents or biomass. Some of the oxygen in the air is consumed; the remainder is vented.

If a bioreactor malfunctions and produces off-specification effluent with a high organic chemical content, the affected bioreactor module (consisting of four ICB units) is isolated from the rest of the system and operated in batch mode until the problem is resolved.

The ICBs are expected to produce the following products:

- carbon dioxide, water, and biomass (solid products of the biological cell mass produced in the reactions; they consist of microbial organisms, residues from organisms, adsorbed metals, grit, and dirt)[1]
- other products, such as chloride and sulfate salts
- some low-molecular-weight, partially oxidized species (e.g., acetic acid), as well as some organic compounds that color the aqueous solution (color bodies)

Until the EDS bioreactor test results became available, the design called for sending the liquid effluent from the

---

[1]The technology provider estimates that about 80 percent of the carbon in the process feed is oxidized to carbon dioxide; the balance is either in the organic biomass (sludge) remaining in solution or in the air stripped from the bioreactor. Test results indicate that the amount of organics that are air-stripped is small, on the order of a few kilograms per year projected for the full-scale plant (Parsons, 2000e).

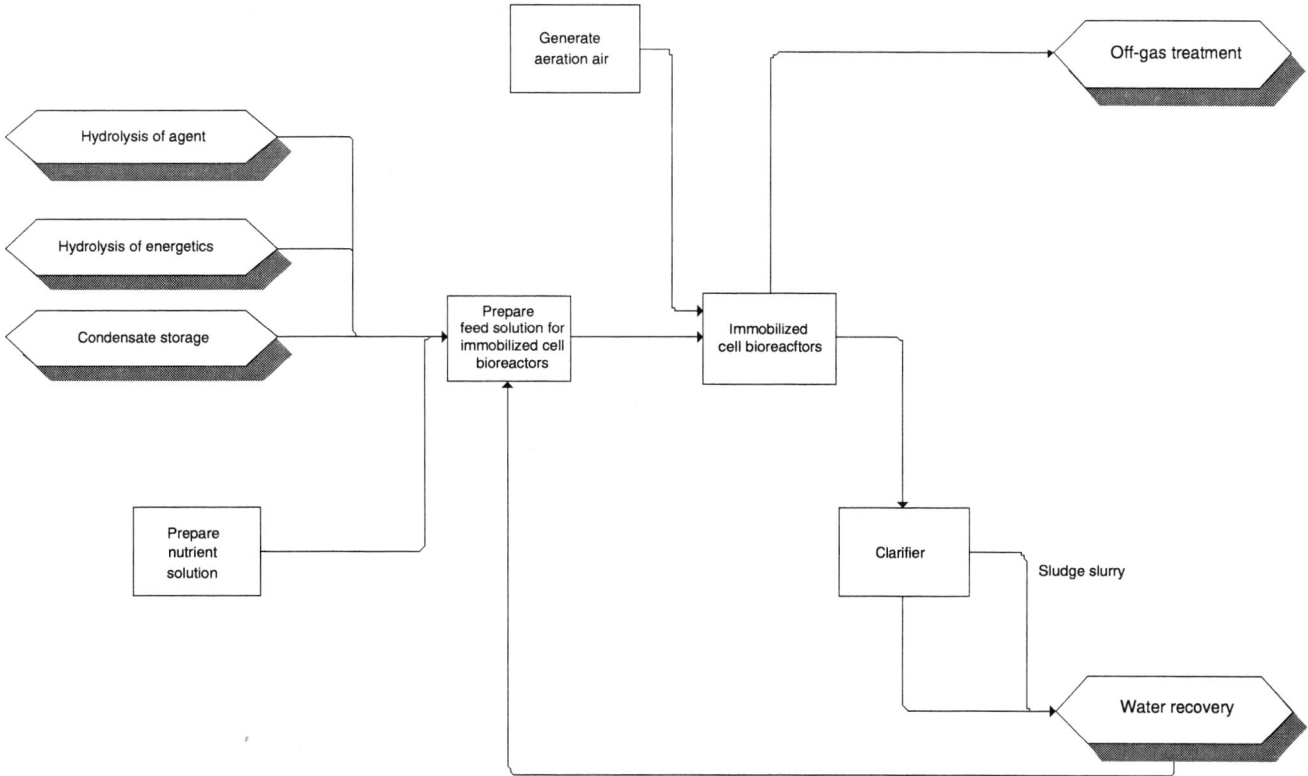

FIGURE 4-4 Biotreatment process. Source: Adapted from Parsons, 2000a.

ICBs to a clarifier for separation into sludge and overflow streams (Parsons, 2000e). The sludge was to be dewatered in filter presses and sent off site to a landfill. The filtrate from the filtration step was to be combined with the clarifier overflow, and the combined stream (about 100 gallons/min) was to be sent to a brine evaporator. The distillate, about 90 percent of the feed, was to be recycled as process water. The bottoms were to be sent to an evaporator/crystallizer for additional water recovery and the crystallized salts sent off site for disposal; the distillate was to be added to the recycled process-water stream. However, EDS test results showed that (1) the clarifier is not needed, (2) the bioreactor effluent can be recycled without clarification, and (3) a slipstream can be sent to the evaporator for removal of salts and sludge. Vented air from the ICBs can be sent to the off-gas treatment system (Parsons, 2000e).

Metal Parts Treaters

Batch Metal Parts Treater

There are two MPTs in the Parson/Honeywell WHEAT design: a batch MPT and a rotary MPT. The batch MPT processes deactivated fuzes, booster cups, nose-closure cups, lifting lugs, and miscellaneous parts from the ERD, as well as burster tubes from the BWM and burster wells from the WMDM. Treatment is done in a batch mode. Munitions parts are placed in containers in a cylindrical vessel heated by external electric-induction coils. The interior of the vessel is swept with superheated steam at slightly below atmospheric pressure. The system is designed to accept munition trays like the ones used at Tooele (Parsons-AlliedSignal, 1999). The munition load is heated primarily by radiation from the vessel walls, with time and temperature conditions designed to meet 5X requirements. Any organic materials present are vaporized or pyrolyzed.

When 5X decontamination requirements have been met, the batch MPT is purged with nitrogen, and following confirmation of the absence of chemical agent in the vapor phase, the tray and its contents are discharged. The tray contents are removed and sent to temporary waste storage or to disposal. The batch MPT vent gas stream is reheated to 677°C (1,250°F) and sent to the MPT quench tower, which the batch MPT shares with the rotary MPT. In the quench tower, the vent gas is contacted with a recirculating alkaline brine solution; the resultant noncondensable vent gases are sent to the MPT off-gas treatment CATOX unit.

A commercial superheater is used to supply steam. The superheater is designed for 50-kW, 15-psig/full vacuum at 538°C (1,000°F) and has a capacity of 138,000 Btu/hr.

Off-gases are heated to 677°C (1,250°F) to ensure complete destruction of the mustard (Parsons, 2000c). This is also done using a standard commercial heater designed for 50-kW, 15-psig/full vacuum at 816°C (1,500°F) and having a capacity of 94,000 Btu/hr and a residence time of 0.5 seconds.

Rotary Metal Parts Treater

The rotary MPT receives munition bodies from the projectile RWM and treats them to a 5X decontamination level. The rotary MPT is cylindrical and contains a rotating internal mechanism with an assembly of cages, baffles, and an internal pipe. Munition bodies are loaded and discharged one at a time through air locks at the feed and discharge ends. As one body is loaded, a treated body is simultaneously discharged. The residence time for each munition body is 75 minutes for 105-mm projectiles and 4.2-inch mortars and 105 minutes for 155-mm projectiles.

The rotary MPT is heated by using external induction coils and swept with superheated steam at near-atmospheric pressure. The wall of the MPT is maintained uniformly at 677°C (1,250°F).

Off-gases from the rotary MPT are passed to a heater that raises the temperature to 677°C (1,250°F) to ensure destruction of any residual mustard. Downstream of this heater, the gas stream is cooled and condensed in the MPT quench tower. A purge stream of the resulting brine is then sent to the MPT/CST condensate holding tanks and from there, if agent free, to the bioreactors. If agent is found, the brine is sent to the agent-hydrolysis system. Noncondensable gases are sent to the CATOX unit of the MPF off-gas treatment system.

Continuous Steam Treater for Dunnage

The CST treats nonprocess waste and dunnage to a 5X decontamination level. Materials fed to the CST include shredded wood pallets, spent activated carbon from the HVAC beds, and shredded plastic from DPE suits.

Wood and DPE suit material are size-reduced prior to being fed to the CST. The activated carbon does not require size reduction. At the time this report was written, the technology provider was planning to use a single, four-shaft shredder with a cutting chamber 44 inches wide by 40 inches long (1,118 mm by 1,016 mm) and powered by a 75 hp electric motor (K. Burchett, Parsons representative, personal communication, November 10, 2000). The shafts and cutters reverse upon amperage overload to minimize jamming. Removable screens are used so the particle size distribution of the product can be changed. The wood chips produced will be about 5 inches long and 0.5 inches wide. A magnetic separator is installed downstream of the shredder to remove metal parts. A similar smaller shredder is used to size-reduce DPE suit material.

The CST is operated in a continuous mode. Its main feature is a horizontally positioned cylinder that feeds materials into one end and moves them through the cylinder by means of a rotating, multibladed auger shaft that turns in a 30-inch-diameter trough running the length of the CST. External induction heating coils are used to heat the wall of the drum and hence all internal materials. Because the interior of the CST has a potentially corrosive environment, especially if condensation forms during cooldown, Parsons/Honeywell has selected Hastelloy C-276 as the material of construction (Parsons, 2000a).

Superheated steam at nominally atmospheric pressure is used as a sweep gas. The amount of steam is 50 percent in excess of the amount needed to destroy the maximum amount of agent preliminarily estimated to be present. The steam is supplied at nominally atmospheric pressure and 538°C (1,000°F) from a superheater, which also supplies steam to the batch MPT (Parsons, 2000a). An off-gas effluent heater heats the vent gas to 649°C (1,200°F) with a residence time of 0.5 seconds to ensure the destruction of any organics present. These gases are then sent to a quench tower, where they are contacted with a recirculating alkaline brine solution. Vent gases from the quench tower are sent to the CATOX unit of the CST off-gas treatment system. A liquid purge from the quench tower is fed to the ICBs.

Granular activated carbon is mixed in with the CST feed, except when carbon from the HVAC system is being treated, to maintain scouring action in the CST and minimize charring during processing runs for DPE suit material. The CST discharge is screened to separate activated carbon from the ash, and the reusable carbon is recycled to the feed end.

Treatment of Off-gases and Disposal of Wastes

The off-gas treatment system has six trains, each with its own CATOX unit (Parsons, 2000a). The monolithic catalyst beds, heaters, reactors, and control systems for each train are of conventional design (Parsons, 2000d). Four trains, one for each ICB module, serve only air vented from the ICB modules. The other two trains serve the MPTs and the CST. Figure 4-5 is a flow diagram of the off-gas treatment system.

Each airstream from an ICB module is heated to between 425 and 450°C (797 and 842°F) and passed to a CATOX unit for the removal of trace organics and oxidizable nitrogen-containing and chlorine-containing compounds before discharge to the atmosphere. The released effluent contains nitrogen oxides, hydrogen chloride, and dioxins and furans below the levels of regulatory concern. The inlet temperature can be lowered to about 371°C (700°F) if upstream process conditions impose a heavier than anticipated organic load. To avoid premature deactivation of the catalyst, the maximum sustained operating temperature at the discharge

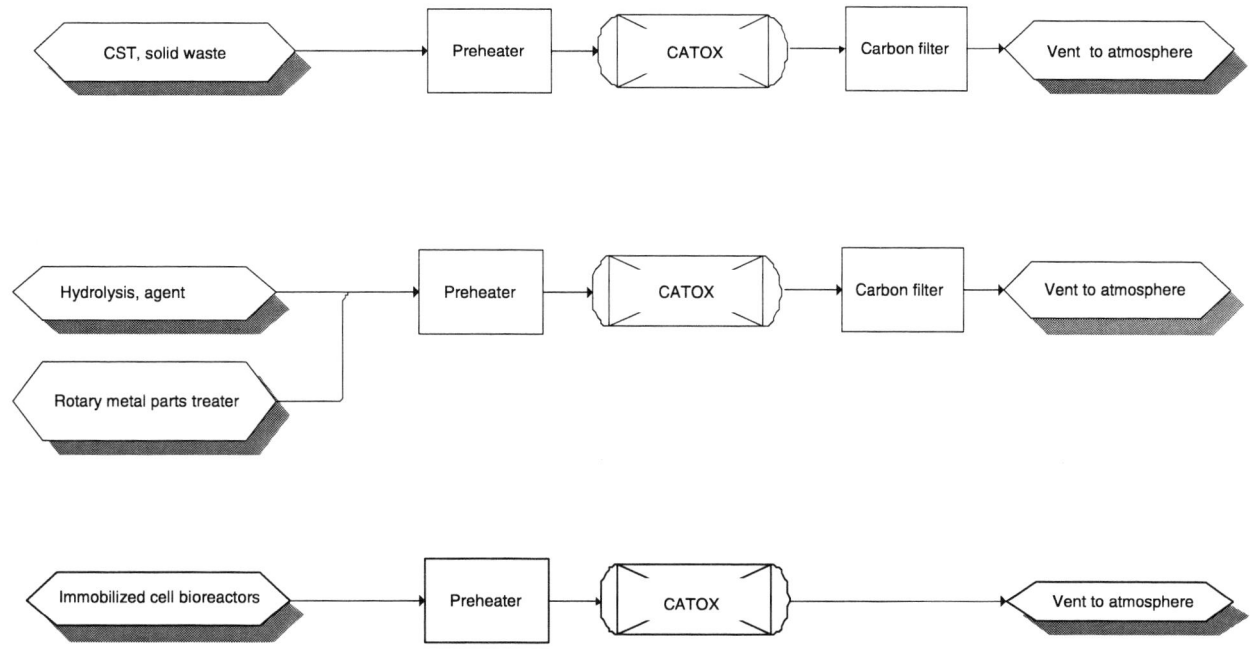

FIGURE 4-5 Off-gas treatment systems. Source: Adapted from Parsons, 2000a.

of the catalyst bed is 677°C (1,050°F). Thus, the systems should be able to handle combustible loads with a rise in adiabatic temperature of 194°C (350°F). Methylene chloride is present in this stream and is only partially oxidized (39 percent) (Parsons, 2000e). At the time this report was being prepared, Parsons/Honeywell was not planning to pass discharge gas from the CATOX unit through activated carbon (K. Burchett, Parsons representative, personal communication, March 2, 2001).

Process gases vented from the rotary MPT, batch MPT, ERD, and various process tanks are sent to the MPT quench tower, from which they are passed in series through a flame arrestor, a preheater, a CATOX unit, and a water-cooled heat exchanger. The cooled gases are then sent to the MDB ventilation system, which contains activated carbon adsorbers. The CST off-gas treatment system has the same design and capacity as the MPT off-gas system.

Liquid Effluent

The Parsons/Honeywell technology package is designed to have no liquid discharges, with the possible exception of a concentrated brine stream from the bioreactor effluent evaporator. All liquid streams are recycled to conserve water, avoid the need for a discharge permit, and mitigate a potential source of public concern. All water introduced into the facility, whether as process water, spent decontamination solution, or water used to wash down equipment, is treated and reused either in the hydrolyzers or at other locations in the facility. However, under some circumstances (e.g., if the humidity is higher than 90 percent), excess water is produced. If so, it is stored for use as makeup water under more typical conditions.

Biosolids

Until the EDS bioreactor test results became available, Parsons/Honeywell planned to separate biosolids from bioreactor effluents by means of a clarifier, followed by dewatering them and compacting them in a filter press. Drummed filter cake would then be sent off site for ultimate disposal in a secure landfill, in the same way as the dried salts (see next section). However, the EDS bioreactor tests showed that this separation step is not necessary (Parsons, 2000e). Therefore, bioreactor effluent containing the biosolids can be sent to the evaporator without removing the biosolids, which remain with the salts. Biosolids and salts are then disposed of together.

Salts

Salts are produced as the result of the hydrolysis of agent and energetic materials, chemical decontamination from washing of the facility, and the biotreatment process. These salts contain metals (e.g., lead) derived from munition components. Dried salts were originally to be crystallized from the brine in an evaporation step. As indicated in the previous section, however, Parsons/Honeywell now plans to remove both biosolids and salts as a single liquid stream during the evaporation step and dispose of them off site.

Metal Parts

Metal parts from the MPTs that have been decontaminated to a 5X condition are subsequently deformed to meet the requirements of the CWC. Historically, this material has been sold to commercial firms as metal scrap.

Nonprocess Wastes

Nonprocess wastes are not direct products of the hydrolysis/biotreatment process but are generated by operational activities or maintenance activities. Nevertheless, this category of material must meet ultimate disposal criteria. Uncontaminated waste is not processed other than being packaged for disposal according to government regulations. Contaminated waste requires processing that renders it suitable for disposal. Contaminated waste must be strictly separated from uncontaminated waste.

Dunnage, which includes packing materials such as wood pallets, fiberboard, steel bands, glass, plastic, and paper, is retained for disposal by a nonincineration means that has not been specified. The Parsons/Honeywell process intends to follow the lead of the baseline system in selecting a nonincineration disposal method (the dunnage incinerator at Tooele has not performed adequately).

Another major nonprocess waste stream is used personal protective clothing and equipment, which includes DPE suits, Tyvek coveralls, gloves, boots, masks, canisters, filters, hoses, and other items. The disposal method for this type of waste depends on whether it is at a 3X decontamination level or has never been in contact with agent. The current baseline method is to retain this waste for placement in a hazardous waste landfill as a listed waste. Parsons/Honeywell proposes to process DPE suits and other personal protective wastes through the CST after size reduction to improve homogeneity and handling. If this does not prove to be practical, the baseline system method for disposal of these wastes would be used.

Other nonprocess waste streams include waste oils and spent hydraulic fluids, which, if contaminated with agent, might be treated in the hydrolyzer and bioreactor systems. Spent activated carbon might be treated in the CST.

Changes to Process

Table 4-1 shows the changes made to the Parsons/Honeywell WHEAT process since the Demonstration I tests (NRC, 2000).

INFORMATION USED IN THE DEVELOPMENT OF THE ASSESSMENT

Engineering Design-Related Documents

To assess the components of the proposed Parsons/Honeywell design, the committee used the following sources of information:

- The 1999 NRC report *Review and Evaluation of Alternative Technologies for Demilitarization of Assembled Chemical Weapons* (NRC, 1999) and the supplementary report *Evaluation of Demonstration Test Results of Alternative Technologies for Demilitarization of Assembled Chemical Weapons* (NRC, 2000), the latter of which focuses on the results of the Demonstration I tests. Since the supplementary report was published, the process has been modified in several respects. Additional testing was under way while the present report was being prepared.
- Documents received at a preliminary EDS review at Parsons on August 2 and 3, 2000. These include Project Design Note (T) – 002 (Parsons, 2000b) and Project Design Note (T) – 005 (Parsons, 2000f).
- Documents from the EDP received at an EDS review conducted by Parsons Infrastructure and Technology

TABLE 4-1 Changes to the Parsons/Honeywell Process Since Demonstration I

Process Area	Former Configuration	Current Configuration
Munitions disassembly	No control for effervescent spillage of mustard.	Cylindrical splash guard to control effervescent spillage of mustard.
Munitions disassembly	Baseline agent-drainage system.	Tilt-and-drain station for agent drainage.
Munitions disassembly	No removal of solidified agent from munitions.	New rotary washout machine.
Biotreatment	Effluent oxidized with Fenton's reagent.	No oxidation step.
Biotreatment	Partial recycling of clarifier effluent.	Most bioreactor effluent recycled without treatment. No clarifer. Remaining effluent processed through brine concentrator and evaporator/crystallizer.
Biotreatment	Salt removed with rotary-drum dryer.	Salt (and biosolids) removed with evaporator/crystallizer.

Group on November 8, 9, and 10, 2000. These include copies of the preliminary hazard analysis (Parsons, 2000c), visual aids (Parsons, 2000a), the design basis manual (Parsons, 2000d), and process and block flow diagrams (Parsons, 2000a).
- Handouts from Scott Susman, PMACWA engineer, at briefings on the status of the EDS on August 9 and October 19, 2000 (Susman, 2000a, 2000b).
- Results of tests of the ICB system (Parsons, 2000e) and of a catalytic oxidation unit (Parsons, 2000g).

Engineering Design Studies Tests

Four series of tests were scheduled in support of the Parsons/Honeywell EDS: (1) tests on the biotreatment, evaporation, crystallization, and filtration steps for the treatment of mustard agent hydrolysate (Parsons, 2000a, 2000e), (2) tests by IITRI, in which mustard agent was fed to a CATOX unit during 786 hours of operation (Parsons, 2000g), (3) CST tests on dunnage (in progress as this report was being drafted) using granular activated carbon (Parsons reported that low levels of dioxins and furans were formed in the CST during processing of DPE suits [K. Burchett, Parsons representative, personal communication, December 27, 2000]), and (4) tests of the projectile washout concept using a small-scale agent hydrolysis reactor, a scaled-down munition MPT, and a CATOX unit (not yet carried out at the time this report was being prepared).

ASSESSMENT OF PROCESS COMPONENT DESIGN

Disassembly of Munitions and Removal of Agent and Energetics

Transportation to Pueblo

Munitions are transported from the depot storage igloos to the Pueblo disposal facility in MAVs rather than in on-site containers on a flatbed truck. The decision to use MAVs, even though flatbed trucks are used at Tooele, is based on the much lower toxicity of mustard agent compared with the GB and VX nerve agents. A MAV is a sealed van, the air space of which can be sampled to ensure that no munitions have leaked during transport. The use of the MAV is being validated by an ongoing transportation risk assessment. Pueblo already has a MAV, and a MAV is being used to transport munitions at CAMDS.

Projectile/Mortar Disassembly

The steps for removing fuzes, bursters, and miscellaneous parts in the WPMD are the same as in the baseline system PMD. The components that transport the disassembled parts to their destinations differ: (1) nose plugs, fuzes, and miscellaneous parts are moved by conveyor to the ERD, (2) bursters are moved to the BWM by a pick-and-place machine, and (3) projectiles/mortars without bursters are moved to the WMDM by conveyor (the same as in the baseline system). The changes from the baseline PMD should not unduly affect the performance or safety of the WPMD.

Burster Washout Machine

The BWM has a rotary carousel with multiple burster-holding receptacles; 12,000-psi water-jet probes (no abrasive) are used to wash out the burster tubes. Water-jet washout of the M55 rocket burster was successfully demonstrated during the ACWA Demonstration I tests using 12,000-psi water.[2] The use of water-jet technology at a Pueblo disposal facility should be straightforward. Pressure should not build up within the burster tubes because the jet direction is tangential toward the walls and the tubes are open at both ends (K. Burchett, Parsons representative, personal communication, February 1, 2001). Lower pressure jets, 2,000 to 3,000 psi, may be used in full-scale operations.

The washout water and the accompanying energetics particles are directed to a shredder to reduce the particles to <0.25 inch in diameter. The technology provider believes that enough water can be provided during shredding to prevent the ignition of the energetics. The BWM has not yet been built, much less tested, and the potential for separating larger energetics particles from the washout solution has not been evaluated. Shredding of the particles without sufficient washout solution could result in ignition. If larger energetics particles cannot be separated from the bulk washout solution, Parsons should consider design modifications to prevent or mitigate ignition.

Energetic Rotary Deactivator

The ERD was not tested during the Demonstration I or EDS phases of the ACWA program. The unit operation receives fuzes and burster cups from the munition disassembly process. The fuzes of artillery rounds contain an AN #6 priming mix, lead azide, tetryl, and black powder. The burster cups contain lead azide and tetryl. The ignition temperature ranges of these materials listed in the design basis manual are incorrect (Parsons, 2000d). Based on the onset of the exotherm in data from differential scanning calorimetry in the explosives handbook from the Lawrence Livermore National Laboratory (Dobratz, 1981), these numbers, with the exception of TNT, are incorrectly listed as degrees Fahrenheit (°F) rather than degrees Celsius (°C). Parsons plans to operate the ERD with a shell temperature of 677°C (1,250°F) and estimates that a minimum 5-minute residence time will be necessary to heat the fuzes and burster

---

[2]A review of water-jet technology for demilitarizing ordnance can be found in Appendix G of the ACW I Committee's report (NRC, 1999).

cups to the operating temperature of 343°C (650°F). This temperature is only marginally above the ignition temperature of lead azide and may therefore be too low to ensure its thermal deactivation. The other energetic materials will be thermally deactivated in the ERD.

The ERD is designed to contain the overpressure of the energetic reaction caused by exposure of the energetic materials in the fuzes and burster cups to the elevated temperature. A nitrogen atmosphere is maintained in the ERD by a constant purge flow of nitrogen gas to ensure that only thermal deactivation reactions occur. The feed rate to the ERD must be controlled to limit the amount and type of energetic materials in the ERD at any one time and thereby avoid exceeding ERD overpressure design limits.

Multipurpose Demilitarization Machine and Projectile Rotary Washout Machine

The Parsons/Honeywell and baseline system MDMs have the same purpose, to access and remove agent in unbursted munitions. The WMDM differs from the baseline MDM in several ways: (1) the original baseline MDM configuration had no way to contain effervescent agent spillage, (2) the WMDM includes a step to open the 4.2-inch mortar for accessing solidified agent (see below), (3) the WMDM uses a tilt-and-drain station instead of a suction tube, and (4) solidified agent is removed in the projectile RWM, which has no equivalent in the baseline system.

The WMDM and RWM were scheduled to be tested with 86 mortars in early 2001 with a small-scale agent hydrolysis reactor, a scaled-down MPT, and a CATOX unit. Removal of solidified agent from mortars is expected to be the most difficult part of the operation because of the internal baffle structure. These early tests will focus on the best way to open the mortar for washout. Reverse soldering and mechanical cutting of either end of the mortar will be tested. Several of the access methods will be tested to determine washout parameters. Another test objective is to observe phase separation to reduce the amount of water that goes to the agent storage tanks. The chemical and physical characteristics of drained agent and washout solids will also be determined.

Hydrolysis of Agent

Toxic Cubicle

Drained agent from the WMDM, agent concentrate from the RWM, MPT condensate, CST condensate, and spent decontamination solution from the agent and energetics hydrolyzers are stored in three cubicles in the MDB. The tanks that hold agent and agent concentrate are located in the toxic cubicle. Other streams that might contain agent are collected in the other two cubicles. The committee was not convinced that adequate instrumentation has been provided to monitor for leaks of agent in the toxic cubicle.

Agent Hydrolysis Reactors

Six reactors hydrolyze agent drained from the WMDM, the agent concentrate from the projectile RWMs, and agent in spent decontamination solutions. The hydrolyzers are also used to destroy any agent detected in the MPT/CST condensate.

Agent from the holding tank in the toxic cubicle (drained agent) is pumped, along with hot water, through a static mixer eductor, which disperses the agent in the water. The agent-in-water dispersion is pumped to a well-stirred 2,520-gallon (1,525-gallon working capacity) PVDF-lined, hot-water-jacketed reactor partially filled with hot water at 90°C (194°F). If modified titanium, which is being considered, is the material of construction for the reactor, ferric chloride will be added to the reactant mass to maintain an oxidizing environment. The initial agent concentration in the reactor is approximately 4 percent.

The hydrolysis of mustard agent has been studied extensively (NRC, 1999, Appendix D). The reaction is mass-transfer-controlled, and vigorous agitation accelerates the reaction. Mustard agent in stored munitions may be only 80 percent pure. As a consequence, although thiodiglycol is the primary product, many other compounds may be present in the hydrolysis product that must be taken into account in downstream operations. Hydrolysis will be the disposal method for mustard agent stored (in bulk only) at the Aberdeen, Maryland, storage site. If the design for Pueblo takes into account the existing knowledge base and lessons learned from ongoing studies of disposal of the Aberdeen stockpile, operational difficulties at Pueblo should be minimal.

Hydrolysis of Energetics

Three batch reactors, each with a nominal 300-gallon capacity and a 200-gallon working capacity, will be used to hydrolyze energetics, including TNT, tetryl, tetrytol, and propellants from mortars and 105-mm projectiles. Propellants will be processed during the final disposal campaigns prior to closure of the facility. The process is designed for 8-hour batches, with two reactors operating and one in reserve. The reactors are constructed from 304 stainless steel and are designed for 150 psig and full vacuum at 121°C (250°F). The reactor temperature of 90°C (194°F) is controlled with external heating and cooling jackets.

The reactors are charged with a 50 weight percent NaOH solution and hot process water prior to the addition of the energetics slurry, which is 20 weight percent energetics. The slurry is generated from the washout water and solid energetic materials removed from the munitions in the BWM. Prior to introduction into the hydrolysis reactors, the slurry mixture from the BWM is shredded in the energetics shredder to ensure the proper particle-size distribution.

The gases from the reactor are sent to the MPT quench

system. A sample test stream from a recirculation loop is analyzed to determine the extent of reaction. This circulation loop also increases agitation in the reactor. The final energetic material load in the reactor is designed to be 12 weight percent after all reactants have been added. The caustic-to-energetics molar ratio for TNT or tetryl is designed to be 3:1 (Parsons, 2000a).

The Parsons/Honeywell design team does not plan to test the energetics hydrolysis system during the EDS phase. The design is based on tests done at the Pantex Plant in Amarillo, Texas, in collaboration with LANL. Energetics hydrolysate for the Demonstration I testing was provided from the Pantex Plant. The major design differences between the EDS Parsons/Honeywell energetics neutralization reactor system and the Pantex Plant system are as follows:

- The energetics are fed as a slurry in the Parsons/Honeywell process, whereas they are fed as solid feed at Pantex.
- The material of construction is 304 stainless steel for the Parsons/Honeywell process; the reactor at Pantex is glass-lined.
- The hydrolysate will not be filtered in the Parsons/Honeywell process as it normally is in the Pantex process.

Biological Treatment

Parsons/Honeywell proposes using modules of parallel ICB units to biodegrade the organic constituents in agent and energetics hydrolysates and MPT/CST condensate, followed by catalytic oxidation of gaseous effluents and recovery of water and dried solids from liquid effluents by evaporation/crystallization. This process raises several issues that must be addressed.

The efficacy of the biological treatment of thiodiglycol and related organic constituents in mustard hydrolysate has been well established at both bench scale and pilot scale (NRC, 1993, 1996, 1999). The biotreatment of a mixture of mustard and energetics hydrolysates has been successfully demonstrated (NRC, 2000). Biological oxidation under aerobic conditions is accomplished by nonspecific mixed microbial populations in a variety of reactor configurations, including conventional industrial wastewater treatment facilities. Results of the EDS tests confirmed that this technology is effective in treating the planned feed stream to the ICB modules (i.e., a mixture of mustard and energetics hydrolysates and MPT/CST condensates) (Parsons, 2000e). Thiodiglycol was not detected in process effluents from the ICBs during these tests.

Nevertheless, the current design, which calls for intermediate storage of mixtures of MPT/CST condensate and agent hydrolysate, should be reconsidered. The composition of the MPT/CST condensate will probably vary because of the variety of waste streams treated in the MPT/CST. A change in the composition could result in poor treatment or even upsets in the ICBs. If MPT condensate and CST condensate are stored separately, adjustments of the feed rate to the ICBs could be made as necessary.

Currently, the biological treatment step (i.e., ICB units) appears to be the rate-limiting step for overall throughput. The bioreactors are designed to treat up to 22 batches of agent hydrolysate per day; the maximum theoretical treatment of agent is 36 batches per day at 4 percent agent loading into the hydrolysis reactor. These parameters suggest that plans to increase the agent loading to 8 or even 12 weight percent in the hydrolysis reactor could adversely affect throughput for the biological treatment of hydrolysate. In addition, biological processing of washout and decontamination streams at higher rates than the design rates will be difficult to achieve. Thus, the overall destruction rate of the Pueblo stockpile will be limited by ICB throughput capacity.

The ICB units must be kept in operation independent of upstream processing, and operating conditions must be maintained for maximum TOC (total organic carbon) loading and minimum hydraulic retention time, while maintaining treatment effectiveness. These objectives might be realized by decoupling the ICB units from upstream processing by increasing the intermediate storage capacity for agent and energetics hydrolysates, and by incorporating off-site treatment of hydrolysates into the facility design. Preliminary EDS results suggest that most of the biological treatment in the ICBs occurs in the first stage; limited treatment occurs in the second stage; very little TOC removal occurs in the third stage. Therefore, increasing the treatment rates of the ICBs might be accomplished by increasing the TOC concentration of the Stage 1 inlet feed, reducing hydraulic residence time, or providing additional hydrolysate feed to the second stage.

Simplification of the water-recovery process sequence following treatment in the ICB units would improve process operations and reduce capital expenditures. Industrial experience and preliminary testing indicate that solids in the crystallizer could be cohesive or sticky, that foaming could occur, and that environments could be extremely corrosive. These conditions could exist under both acidic and alkaline conditions because of the presence of chlorides, sulfates, phosphates, nitrates, and carbonates.

Based on preliminary EDS results indicating low total suspended solids in ICB effluent, the sludge clarifier/thickener might be eliminated. Off-site disposal of concentrated brine rather than dried salts would eliminate the need for the crystallizer and solids filtration steps. Off-site brine disposal would also be consistent with current operations at Tooele, Utah, and planned operations at Newport, Indiana. The basis for the currently designated materials of construction for the evaporator and crystallizer is unclear. If on-site crystallization is included in the final process design, a more extensive evaluation of the materials of construction will be necessary, including stressed-materials coupon-exposure tests. The dried solids from the crystallization step might be considered a RCRA hazardous waste because of their origin or because of the potential for a highly variable heavy metals content (owing to variability in mercury or other heavy metals in the munitions).

The CATOX unit for the gaseous effluents from the ICBs should be designed to withstand corrosion and a high rate of scaling from entrained liquids. Entrained liquids are likely to contain chlorides, nitrates, sulfates, and phosphates under mildly acidic to alkaline conditions (pH 6 to 9). Preliminary results from the EDS investigations suggest that localized scaling and corrosion are likely.

Metal Parts Treatment

Batch Metal Parts Treater

The batch MPT processes deactivate fuzes, booster cups, nose-closure cups, lifting lugs, and miscellaneous parts from the ERD, as well as burster tubes from the BWM and burster wells from the WMDM. The interior of the vessel is swept with superheated steam at slightly below atmospheric pressure from a superheater with time and temperature conditions designed to meet the 5X decontamination requirements. Any organic materials are vaporized or pyrolyzed. The batch MPT is purged with nitrogen between batches. Following confirmation that no chemical agent is present in the vapor phase, the tray is discharged and its contents sent to temporary storage or final disposal.

A potential problem is the leakage of air into the batch MPT, which could cause a fire. The PHA addresses this issue for two failure types: (1) breaches of doors or walls or failure of the air-lock door seals caused by high temperature and (2) loss of nitrogen purge. A breach or seal failure would cause a release of steam and toxic/flammable gas into the room and fire in the room. The fire-suppression system would mitigate the effects of fire and agent release. The response to the loss of nitrogen purge at the end of the treatment with steam would be a shutdown of the external heater, triggered by high temperatures in the batch MPT.

Rotary Metal Parts Treater

The rotary MPT receives munition bodies from the projectile RWM and decontaminates these bodies to a 5X condition. The rotary MPT is heated by external induction coils and swept with superheated steam at 12 psia pressure. The wall of the MPT is maintained uniformly at 677°C (1,250°F).

One concern is the potential for jamming of the munitions bodies within the rotary MPT because of the high temperatures (649°C; 1,200°F). The following are some possible causes of jamming:

- warping of the munitions bodies because of thermal stresses
- softening and flowing of munition components
- breaking off or cracking of components with shards or edges that can lodge in the MPT framework
- melting of silver solder and fusing of the munitions to the MPT framework

Another concern is agent leakage from the rotary MPT during maintenance activities or normal operation. Although the munition bodies will have been washed out, at times agent might in fact be present. Air locks (i.e., chambers with inlet and outlet doors) will be provided at both ends of the rotary MPT to prevent leakage of air into the rotary MPT and leakage of agent out of the rotary MPT. Consequently, the effectiveness of the air locks is critical. The possibility of leaks into and out of the rotary MPT is addressed in the PHA. Negative pressure in the rotary MPT, positive pressure (saturated steam) in the air locks, and the presence of a fire-suppression system, as well as other effective controls, are recommended in the PHA.

Treatment of Dunnage in the Continuous Steam Treater

The CST treats nonprocess waste and dunnage to a 5X decontamination level. Shredded wood pallets, spent activated carbon from the HVAC beds, and shredded plastic from DPE suits are fed to the CST. Wood and DPE suit material are size-reduced by a process that has two four-shaft shredders, one for wood and one for DPE suit material. The shredder for wood has a cutting chamber 44 inches wide by 40 inches long. The shafts and cutters reverse upon amperage overload to minimize jamming. A similar but smaller shredder is used to size-reduce DPE suit material.

Even with the automatic reversing feature, jamming of the shredder could occur, and lodging of items in the feed chute should be anticipated. Therefore, the feed chute to the shredder must be large enough to ensure that the largest feed item (probably a pallet) cannot become trapped in the chute. Tests must be run with chutes and shredders of the same dimensions as those planned for full-scale operation to ensure that the design of the feed chute is adequate.

Selection of the materials of construction for the Parsons/Honeywell CST will be critical. Hydrogen chloride, a very corrosive material, will be produced when DPE suits are treated. Therefore, caustic (form not specified) will be added to neutralize the hydrogen chloride. Condensate may form when the CST is shut down.

The committee expects that combustible gases (carbon monoxide and hydrogen) will be formed in the CST as a result of the reaction between steam and activated carbon. The formation of these gases would reduce the amount of steam available for the hydrolysis of mustard; in addition, the gas formed will be highly flammable. This possibility has apparently not been explored to date.

The PHA addresses the potential problem of leakage of air into the CST, which could result in a fire or other unwanted chemical reactions. The cooling-water temperature will be monitored to ensure that the temperature remains within design limits. If it exceeds the limits, the external heater will be shut down. Leakage of gases out of the CST will be prevented by negative pressure in the vessel, which must be monitored. A fire-suppression system is also prescribed.

Off-gas Treatment and Disposal of Wastes

The Parsons/Honeywell process features three separate off-gas treatment systems with three separate CATOX units. The CATOX units for the ICBs will discharge off-gas directly to the atmosphere; the units for the MPTs and CST will discharge off-gas to the MDB carbon filters as a precaution against venting any unreacted agent. Because the hydrolysates fed to the ICBs are tested to ensure that they are free of agent, it is reasonable to assume that carbon filters will not be necessary to capture unconverted agent from the ICB-fed CATOX units. However, trace pollutants could be generated by the oxidation of volatile organic compounds in the ICB vent gas. Also, Parsons indicated in a preliminary report that low levels of dioxins and furans, believed to originate in the CST, were present in the off-gases from the bioreactor CATOX units. At the time this report was being written, preliminary data from the EDS tests indicated that residual pollutants from the CATOX units treating off-gas from the ICBs were below EPA regulatory limits (Parsons, 2000e).

The Parsons/Honeywell EDS also included tests of CATOX units to ascertain if residual mustard agent that might have survived steam treatment in the MPTs and CST had been destroyed (Parsons, 2000a). These units were challenged with 10 mg/m^3 of mustard agent, compared with expected values of less than 3 $\mu g/m^3$ in a full-scale operation. The test results showed that the CATOX units could successfully achieve a destruction and removal efficiency of 99.999 percent at this level for a period of 637 hours. No catalyst deactivation or increased pressure drop was observed. However, there was some plugging of the effluent cooler, presumably by products of combustion of the mustard agent.

Because levels of mustard agent in the off-gas from the MPTs and CST in a full-scale facility are expected to be orders of magnitude lower than the 10 mg/m^3 used during testing, the presence of mustard agent combustion products in the effluent from the CATOX units would be rare. However, a significant level of organic compounds could be present in the effluent from the CST. The 538°C (1,000°F) steam used for the treatment of dunnage in the CST could produce a variety of pyrolysis products. After post-treatment of these products with steam at 1,200°F, a complex mixture of combustible compounds could still exist in the gas phase after the quench tower. One could also envision the formation of a smoke or soot aerosol in the CST quench tower on cooling that would be difficult to scrub out. To date, the CST vent-gas stream has not been characterized (with the exception of the detection of low levels of dioxins and furans) or tested with a CATOX unit to see what remains after treatment. Carbon filters may remove residual pollutants, but this has not been demonstrated.

The Parson/Honeywell technology package is designed to produce no liquid discharges, with the possible exception of a concentrated brine stream from the bioreactor effluent evaporator. Careful consideration has been given to the disposition of this brine and all other waste streams.

ASSESSMENT OF INTEGRATION ISSUES

Component Integration

Destruction of the Pueblo stockpile within the time specified by the CWC treaty requires that the overall process achieve the required throughput levels and process availability (i.e., the fraction of time the plant can operate). Thus, the throughput and availability of each process step, in combination with equipment redundancy and sufficient buffer storage capacity between process steps, must result in the specified destruction rate. Integrating the individual processing steps will require effective process monitoring and control to ensure that appropriate materials are fed at each step and that all materials discharged from the plant meet safety and environmental specifications. In addition, attainment of the required process availability depends on the durability of all materials of construction and the effectiveness of the plant operating and maintenance force.

The Parsons/Honeywell WHEAT technology package is a combination of continuous and batch processing steps with selective buffering capabilities between some of the processing steps. Parsons has designed the process and specified the size and number of components to enable the disposal of the entire stockpile at Pueblo in 29 months; this will entail operating the plant 24 hours per day, 7 days per week, and an overall plant availability of 52 percent (Parsons, 2000d). Each type of munition will be handled in a separate disposal campaign.

Process Operability

The maximum integrated plant capacity is based on the operation of the reverse-assembly and agent-accessing systems (the WPMD/WMDM combination) for 63 percent of the time, or about 15 hours per day; downstream agent-neutralization processing systems must operate 24 hours per day, using all of the installed system's capacity. The design feature that is expected to limit capacity of the integrated facility is the capacity of the bioreactor system, that is, the number of ICB units. This limitation could be eliminated by installing more ICB units, but at increased cost. The number of individual components and systems has been chosen to balance costs and expected average throughput rates and on-stream time. An alternative would be to increase the concentration of hydrolysates in the feed. However, the ICB system may not adequately destroy the components of the hydrolysate if the concentration is increased. Potential impacts of increased concentrations have been previously discussed.

Materials of construction for the plant equipment have been selected to provide reliable operation, so that the Pueblo stockpile can be disposed of by the CWC treaty deadline. If the EDS test results indicate that the composition of feed streams will damage the equipment, more resistant materials of construction will be specified. Moreover, further refinements in the process are expected to result in changes in materials of construction that will be incorporated into the next developmental engineering phase for the WHEAT technology package.

Monitoring and Control Strategy

The monitoring and control strategy is based primarily on existing methods and systems in use or planned for use at other CSDP facilities or similar commercial installations (e.g., industrial wastewater treatment plants). All of the instruments and control elements are standard industrial hardware with field-proven high reliability and robustness. The overall system consists of the BPCS, ESS, and individual equipment PLCs. The BPCS contains microprocessor-based controllers. The ESS is a separate, dedicated safety system consisting of PLCs or microprocessor-based controllers that will provide protective logic and enable safe shutdown of the facility.

Process Safety

Energetic disassembly, burster tube washout, and fuze deactivation take place in explosion containment areas. The hydrolysis process operates at up to 90°C (194°F); the bioreactor (ICB) processes operate at ambient temperature. Both processes operate at ambient pressure. Handling failed burster pulls at the WMDM cutting station and accessing residual agent for washout by the projectile RWM involve metal-cutting operations that have not been defined at this stage, but the associated temperature, pressure, and explosion hazards are expected to be in an acceptable range. The CATOX units operate at 425°C (797°F); the MPTs operate at up to 677°C (1,250°F). The hydrolysis reactors and the MPTs, which represent the primary detoxifying processes, operate either in a batch mode or, in the case of the rotary MPT, have batch-type monitoring stations. The energetics hydrolysis reactors also operate in batch mode. Thus, the effectiveness of treatment will be ascertainable prior to the release of material to the next step.

The remaining systems are routine chemical processes and occur downstream of the primary (hydrolysis units, MPTs, and the CST) and secondary (biotreatment) detoxifying processes. These systems, which include equipment such as scrubbers, carbon filters, evaporators, and crystallizers, should pose no unique hazards.

Worker Health and Safety

The Parsons/Honeywell EDS design includes the removal of propellant from 105-mm projectiles and the propellant and ignition cartridge from 4.2-inch mortars in the PRR adjacent to the UPA (most of the 105-mm projectiles and 4.2-inch mortars do not contain propellant). In the design previously evaluated in the ACW I Committee report, this operation was to be performed remotely (NRC, 1999). The hazards to workers in the PRR are similar to those in the UPA, and the consequences of accidents are also largely the same. In the new design, munitions are handled by workers more often than in the previous design, which could increase the potential for accidents.

Experience at JACADS and Tooele indicates that the handling of munitions is a low-risk operation. The munition (inside its fiberglass container) is placed in a glove box and monitored for the presence of agent prior to its removal from the container. Thereafter, the munition is handled for propellant removal in the open (i.e., not in a special ventilation enclosure). Before total-solution operations for the Parsons/Honeywell technology package can begin at Pueblo, a QRA will be performed to ensure that risks in the PRR are low.

Following removal of the propellant, the munitions are moved to the ECR, where remaining energetic materials are separated from the munition body. The remote separation of energetics and agent in a facility designed to contain explosive overpressure reduces worker hazards. Disassembly operations to separate energetics and accessing and draining agent are done by machine (WPMD and WMDM), robot arms (including pick-and-place machines), and conveyor systems similar to baseline system processes and are not expected to introduce new hazards or to increase hazard levels over similar baseline operations. The BWM incorporates an energetics shredder, which has raised concerns about the potential for energetics ignition. Metal-cutting steps will be necessary if the WMDM fails to remove a burster well and when accessing agent cavities for washout in the projectile RWM. Both operations are performed remotely and present hazards to maintenance workers only if they enter these areas for maintenance work. The committee expects that appropriate safety precautions (e.g., lock-out and tag-out, job hazard analysis, and training) will be taken to protect workers.

The current Parsons/Honeywell WHEAT design introduces complex machinery (e.g., the projectile RWM) that is not included in the baseline system or in the earlier WHEAT design. Perhaps the most significant worker safety issue is maintenance of this specialized equipment in DPE suits, particularly during start-up and early disposal campaigns. Maintenance requirements for complex equipment that could contain energetic residues in the presence of other equipment with sharp edges have been increased throughout the

EDS phase of WHEAT design. Maintenance can be performed safely but will require increased attention to safety as the design progresses.

The ERH may have an external recirculation loop. Pumping an aqueous slurry of energetic materials can be done safely under the proper conditions. If an accident occurs during normal operations, there would be little risk to workers because they are not expected to be present. The recirculation loop and other components of the energetics hydrolysis system should be designed to ensure that energetic material cannot accumulate in the piping, valves, or pumps. There is an ongoing research program at HAAP to define safe design and operating parameters for the ERH reactor. Results of this program should be carefully considered in the final design of the WHEAT hydrolysis reactor.

If a process upset occurs requiring an emergency shutdown, the products of incomplete hydrolysis in the agent and energetics hydrolysis reactors will be extremely hazardous. The ERD, rotary MPT, batch MPT, CST, and CATOX units will be decontaminated through a time-at-temperature process, but a forced shutdown might not result in complete decontamination. Design features and procedures are expected to be established for safe shutdown, restarting, and/or maintenance of the systems that precede these units in the event of a forced shutdown. The air effluent during an upset would continue to be treated, first in the CATOX units, then in activated carbon filters.

Only trace amounts of energetics will be present in the batch MPT under expected operating conditions. The batch MPT design incorporates an oxygen-free atmosphere and a robust design to accommodate some carryover of energetics from the BWM. Scenarios for the introduction of energetics beyond design conditions, such as a major gross washout failure in the BWM, will be evaluated to ensure that they are extremely unlikely before the design is completed.

The ERD operates with a nitrogen gas atmosphere; the batch MPT uses nitrogen purge gas at the end of each decontamination cycle. Consequently, work areas will require monitoring for oxygen deficiency to preclude asphyxiation during normal and maintenance operations.

Potentially flammable dunnage pyrolysis products are being characterized during EDS testing, and the impact of these and other flammable effluents should be considered as the design develops.

The biosludge produced in the ICBs could contain some pathogenic microorganisms. The potential for worker exposure to these microorganisms is expected to be minimized by appropriate protective gear.

The primary hazardous materials used are sodium hydroxide and sodium hypochlorite. These chemicals are used routinely at many industrial facilities and are not unique to the Parsons/Honeywell process.

Public Safety

The release of agent and other regulated substances in effluents from the Parsons/Honeywell process is extremely unlikely. The destruction of agent and energetics is verified by hold-test-release operations before the transfer of hydrolysate from the hydrolysis reactors to the ICBs and before the transfer of bioreactor sludge to the sludge containerization step. The gaseous effluent from the bioreactors is continuously released through catalytic oxidizers and scrubbers. Thus, no hold-test-release operation is provided for the gaseous effluent stream from the ICBs, but release of agent from this stream is considered extremely unlikely because of the source, the upstream monitoring, and the gas treatment steps.

The most likely cause of a release of agent or other regulated substances that might pose a threat to public health would be an explosion or rupture of a pipe or vessel. Very small releases of agent, which posed no public health risk, occurred at JACADS and Tooele during maintenance operations. Wih the incorporation of lessons learned, as well as a QRA, the likelihood of such occurrences at the conclusion of the design process should be extremely small.

Preliminary Hazards Analysis

A PHA was prepared as part of the Parsons/Honeywell EDP, in accordance with MIL-STD-882C. The PHA is based on a preliminary hazards list that describes unmitigated hazards; the list will be updated as the design process progresses (DOD, 1993). Numerous PHA work sheets were generated, and recommendations were made to reduce risk. The PHA results were reviewed by a multidisciplinary group from the Parsons/Honeywell team, which will track the means of mitigating hazards.

Parsons/Honeywell used the PHA for Tooele as a resource document. For systems that are the same as or similar to those at Tooele, the level of detail in the PHA is high. For systems unique to the Parsons/Honeywell technology provider's package, such as the projectile RWM, the level of detail is low. Maintenance failures are addressed primarily on a generic basis (general work sheets), but a few maintenance failures modes are addressed in process-specific work sheets.

The PHA recommendations can be addressed either by design solutions or procedural/administrative solutions. The committee believes that design solutions should be used whenever possible, because a design solution can be implemented rather easily early in the design process and can more easily be structured to minimize the potential for human error.

The ACW II Committee remains committed to the following recommendation by the ACW I Committee (NRC, 1999):

General Recommendation 3. If a decision is made to move forward with any of these technology packages, health and safety evaluations should progress from qualitative assessments to more quantitative assessments as the process design matures. Quantitative (QRA), health (HRA), and ecological risk assessments should be conducted as soon as is practical. Early initiation of these assessments will allow findings to be implemented with minimal cost and schedule impact.

Human Health and the Environment

Effluent Characterization and Impact

In the absence of a health risk assessment and an environmental risk assessment, a precise statement on the impact of effluents on human health and the environment cannot be made at this time. However, data available to date indicate that the gas flow leaving the plant will meet all EPA regulations. Solid waste streams, including uncontaminated dunnage and metal decontaminated to a 5X level, will be agent-free. The bioreactor effluent evaporator will produce a concentrated brine that may be considered hazardous because of the "derived from" rule. It might even be considered hazardous by characteristic, although this is not likely.[3]

Completeness of Effluent Characterization

The very large gas flow, primarily from the ICBs, will have gone through CATOX units and, possibly, activated carbon filtration (K. Burchett, Parsons representative, personal communication, December 27, 2000). All other gaseous effluent will pass through both CATOX units and activated carbon filters. The gas composition will have to be determined in detail during initial trials. The gas should then be tested routinely for chemical agent, oxygen, carbon dioxide, and carbon monoxide on a real-time basis. It should also be characterized for low concentrations of hazardous materials, such as dioxins.

Biomass and salt residue, separately or combined, will also be tested for toxicity and leachability. Effluents that have been treated to a 5X condition will not require further characterization.

Characterization of many effluents is an objective of EDS testing scheduled for completion in 2001. Some preliminary results are given in the following section.

---

[3]Under RCRA, a waste is declared hazardous if it exhibits certain characteristics or if it is listed as a hazardous waste by characteristic. Certain features of the waste, such as flammability or toxicity, can cause it to become considered hazardous by characteristic. The federal government or a state government can pass legislation declaring that a particular waste is a listed hazardous waste, regardless of its features. If a listed hazardous waste is treated, the residues from treatment are considered to remain hazardous wastes, with the possible exception of materials sent off for recovery and reuse; this is the "derived from" rule.

Effluent Management Strategy

Salts. Dried salt, probably containing some organic materials, will contain sodium salts of fluoride, chloride, sulfate, nitrate, and nitrite. Preliminary information from the EDS tests indicates that this stream is not hazardous by characteristic (Parsons, 2000e).

Biosludge. During demonstration testing, the biosludge appeared to be nonhazardous (NRC, 2000). Preliminary EDS testing results on the bioreactor sludge and brine show that (1) most constituents of the toxicity characteristic leaching procedure (the TCLP, as defined by RCRA regulations) are either nondetectable or are at least one order of magnitude below the regulatory limit and (2) dioxins and furans were either nondetectable or equivalent to levels found in uncontaminated environments (Parsons, 2000a). Testing during initial operation and periodically thereafter will be necessary to determine whether or not wastes are hazardous, as defined by the EPA. If so, disposal in a hazardous-waste landfill may not be possible because of biological activity. Incineration of the waste would be an alternative. If the wastes are not hazardous, they can probably be sent to a municipal solid-waste landfill without threat to human health or the environment.

Gas. Exhaust gas from the CATOX units serving most processes except the ICBs will pass through an activated carbon adsorber. In its review of demonstration testing, the the ACW I Committee concluded as follows: "The gas leaving the CATOX unit had traces of low-molecular-weight materials, which are considered acceptable. Chlorinated dioxins and furans were observed at very low levels in some of the analyses, but these compounds should be adsorbed from the gas by the carbon filter" (NRC, 2000).

The ACW II Committee notes that the EDS design does not have a carbon filter on the effluent from the CATOX units serving the ICBs. Performance of the CATOX units was tested during EDS testing of the ICB and the CST. In addition, CATOX units were tested with direct injection of mustard agent. The gas composition was determined both upstream and downstream of the CATOX unit. Preliminary results from the ICB tests show that dioxins and furans were present in the bioreactor and CATOX unit effluents at up to a few hundred picograms per cubic meter. Analyses will be necessary during initial operation and periodically during operation to confirm the presence or absence of low-molecular-weight hydrocarbons and chlorinated hydrocarbons, oxides of nitrogen, and chlorinated dioxins and furans.

In its review of demonstration testing, the ACW I Committee also observed that during direct injection of mustard agent into the CATOX unit, agent was destroyed at a destruction and removal efficiency of greater than 99.9999 percent (NRC, 2000). Preliminary results from the EDS tests show that the CATOX unit reduces mustard agent to nondetectable levels (about 0.1 of the allowed time-weighted average concentration for worker exposure, or 0.3 µg/m^3 (a destruction and removal efficiency of greater than 99.999 percent for mustard agent injection of about 10 mg/m^3 for more than 600 hours) (Parsons, 2000a). Therefore, the presence of agent in process gas effluents is extremely unlikely.

Metal Parts. Metal parts are cleaned and decontaminated to a 5X condition in the MPTs. The cleaned parts are not expected to pose any threat to human health or the environment.

Off-site Disposal Options

Agent Hydrolysate. Hydrolysate from neutralization of mustard agent with water has been demonstrated to be effectively treated by commercial biological wastewater-treatment facilities. In addition, the mustard agent hydrolysate has been delisted as a hazardous waste in Maryland and has undergone extensive toxicological evaluation showing it to be acceptable for shipment according to U.S. Department of Transportation guidelines (NRC, 1996). Off-site disposal of mustard agent hydrolysate is currently being evaluated for the Aberdeen facility (Myler, 2000). Several commercial facilities have expressed an interest in the treatment and disposal of the hydrolysate. Off-site treatment of mustard hydrolysate would greatly simplify process integration requirements for Pueblo. Hydrolysate produced from mustard agent at greater concentrations than the current design basis of 4 percent and hydrolysate produced from HT would necessitate additional testing to further evaluate this option.

Brine. Parsons/Honeywell is considering the off-site disposal of concentrated brine from the bioreactor effluent evaporator, instead of evaporating the brine to dryness. Brines produced from air-pollution-control processes at Tooele are currently being shipped off-site for disposal by commercial waste-management facilities. The effluent from the treatment of VX nerve agent hydrolysate by SCWO at the Newport site is also planned to be shipped off-site for disposal after being concentrated by evaporation. This material has been delisted as a hazardous waste by the state of Indiana, and at least 16 commercial facilities have been identified that could accept the brine (Wojciechowski, 2000). Off-site management of ICB effluent after evaporation (to recover water) at Pueblo would eliminate the need for a crystallizer and simplify process integration.

Dunnage. Uncontaminated dunnage and agent-free metal will also be disposed of off-site. Off-site disposal of other agent-free waste streams, such as hydrolysates, may be possible, thereby reducing the number or size of treatment steps. Experience at JACADS and Tooele has shown that only a small fraction of dunnage is contaminated with agent. Uncontaminated dunnage from these two stockpile locations is being disposed of off-site by commercial waste-management facilities (McCloskey, 2000; U.S. Army 1998). Off-site management of uncontaminated dunnage is also planned for both the Newport and Aberdeen stockpile locations. Off-site management of dunnage from Pueblo would greatly reduce the on-site process requirements and simplify process integration by reducing the quantity and complexity of the feed to the CST and subsequent treatment of the condensate. Off-site disposal may be possible under Colorado regulations.

Resource Requirements

Three steps will consume large amounts of energy: (1) producing 650°C (1,202°F) steam for the MPTs, (2) heating vent gases to about 425°C (797°F) for the catalytic oxidizers, and (3) evaporating the salt solution from the bioreactor. The resource requirements have been estimated at about 11 MW of electricity, 5,700 lb/day of nitrogen (530,000 standard cubic feet per day), and up to approximately 60,000 gallons/day of makeup water. None of these resource requirements appears to be excessive for a chemical plant.

Environmental Compliance and Permitting

There are no apparent reasons for the combination of technologies in the WHEAT process to lead to unusual permitting or compliance problems.

ASSESSMENT OF OVERARCHING TECHNICAL ISSUES

Steps Required Before Implementation

Materials of Construction

Parsons/Honeywell has begun the selection of optimal materials for construction of the critical process units. However, this process had not been completed when this report was being prepared. Experimental testing on the materials selected must be carried out to verify choices.

Continuous Steam Treater

The efficacy of using activated carbon for the aggregate must be verified, and any detrimental side effects must be identified and taken into account in the design.

Previous Findings and Recommendations

In this section, the findings and recommendations regarding the WHEAT process from the two ACW I Committee reports are reviewed to determine if they are still valid or if they have been addressed by the EDS tests or by information from the evolving WHEAT process design (NRC, 1999, 2000).

Review of Findings from the 1999 Initial ACW I Committee report (NRC, 1999)

Finding PA-1. The biological treatment operation will require further demonstration to prove its ability (1) to handle a variety of feed stocks with reasonable acclimation times between changes, and (2) to achieve high levels of conversion of the Schedule 2 compounds in the hydrolysate. The demonstration will have to last long enough to give confidence in the long-term operation ability of the process.

Testing has shown that the biological treatment operation can handle mustard hydrolysate and CST condensate. Schedule 2 compounds can be completely destroyed. Long-term operation appears to be feasible.

Finding PA-2. The relative effects of biological treatment and air-stripping on the destruction of organic materials in the bioreactor have not been established. This will affect the composition of the off-gas from the bioreactor.

The quantity of organic contaminants removed by air-stripping is small in comparison with the amount destroyed by biological oxidation.

Finding PA-3. The effectiveness of ultraviolet/hydrogen peroxide oxidation in reducing Schedule 2 compounds to an acceptably low level has not been demonstrated.

The ultraviolet/hydrogen peroxide treatment step is no longer a part of the process.

Finding PA-4. The bioreactor has been operated only at very low salt concentrations. Operation at design concentrations has not been demonstrated.

The ICB has been operated at the design concentration.

Finding PA-5. Additional data should be gathered on the effectiveness of the catalytic oxidation system in destroying organic materials in the biotreatment off-gas.

Additional data have been gathered and the degree of destruction of organic compounds appears to be adequate.

Finding PA-6. The sludge from the biological process has not been completely characterized.

The process no longer produces this sludge stream.

Finding PA-7. Even though the evaporation operations involve conventional technologies, they have not been tested for this application.

Testing was under way as this report was being prepared. No results were reviewed.

Finding PA-8. The dried salts from the evaporation operations have not been characterized for leachability and toxicity.

The process has changed, and the production of a dried salt stream is not planned.

Review of Findings and Recommendations from the ACW I Committee Supplemental Report (NRC, 2000)

Finding (Demo I) PA-1. The mustard demonstration tests were very encouraging and showed that the process is ready for the next scale-up.

These tests were previously carried out successfully.

Finding (Demo I) PA-2. The nerve agent demonstration tests had serious problems. However, if the previous tests at the technology provider's laboratory and the results of the demonstration tests are combined, the aggregate results are inconclusive. The reason for the poor demonstration results might be as simple as poor aeration in the bioreactor (see Recommendation PA-1).

No additional tests of the ICB were carried out on nerve agent hydrolysates since nerve agents are not stored at Pueblo.

Recommendation (Demo I) PA-1. Before proceeding to a further scale-up of GB and VX biotreatment processing, the committee recommends that the following steps be taken:

- The biotreatment process should be examined carefully at bench scale to determine the factors that are critical to success.
- An investigation of analytical techniques should be undertaken to provide more reliable process information.

No additional testing has been done on nerve agent hydrolysates. The committee is not aware of any further investigations of analytical techniques since nerve agents are not stored at Pueblo.

New Findings and Recommendations

Finding (Pueblo) PH-1. The Parsons/Honeywell design of the burster washout machine is relatively immature and should be demonstrated before it is incorporated into the final technology package. Items of concern include separation of energetics particles from the bulk washout solution, which could potentially lead to ignition in the shredder.

Finding (Pueblo) PH-2. The maximum fuze and burster-cup temperature of 650°F (343°C) in the energetics rotary deactivator (ERD) might not be high enough to ensure the thermal deactivation of lead-azide-containing fuzes. If the temperature of the materials in the ERD is within the range required to ensure safe thermal deactivation of lead azide, the process should be adequate.

Finding (Pueblo) PH-3. The Parsons/Honeywell WHEAT munitions disassembly machine and the projectile rotary washout machine have not been tested to demonstrate reliable removal of solidified agent to levels sufficiently low for treatment in the metal parts treaters. Testing planned for 2001 may provide data to verify that these machines will be reliable enough to achieve target throughput rates.

Finding (Pueblo) PH-4. Leaks into or out of the batch metal parts treater, rotary metal parts treater, and continuous steam treater were adequately addressed in the preliminary hazards analysis.

Finding (Pueblo) PH-5. The rotary metal parts treater has not been tested; the concept, therefore, remains to be demonstrated.

Finding (Pueblo) PH-6. The use of activated carbon and caustic as additives to the continuous steam treater are unproven. Experimental studies to demonstrate their utility were under way as this report was being prepared.

Finding (Pueblo) PH-7. Dioxins and furans are produced in the continuous steam treater (CST) under certain conditions. Some of these dioxins and furans are eventually fed to the immobilized cell bioreactors via the purge stream of the CST quench tower and are then stripped from the bioreactors and partially destroyed in the catalytic oxidation units. Activated carbon adsorbers might, therefore, be necessary downstream of the bioreactor catalytic oxidation units.

Finding (Pueblo) PH-8. The catalytic oxidation units have been successfully demonstrated to destroy mustard agent, thus providing an extra layer of protection in the event of an upset that results in mustard agent breakthrough to the catalytic oxidation units.

Finding (Pueblo) PH-9. Although the catalytic oxidation units have generally performed well in destroying volatile organic compounds, it has not yet been demonstrated that they can reduce pollutants in the vent gases from the continuous steam treater or metal parts treater to acceptable levels.

Finding (Pueblo) PH-10. No outstanding waste disposal issues remain.

Finding (Pueblo) PH-11. Experience with the baseline system and continuing tests of individual components of the Parsons/Honeywell WHEAT process indicate that successful integration of the process components is feasible. However, the individual steps have not been tested to demonstrate that the necessary throughput rates and online availability levels can be achieved.

Finding (Pueblo) PH-12. A detailed throughput analysis that takes into account intermediate storage capacity has not been carried out. A throughput analysis would verify that planned throughput rates can be achieved.

Finding (Pueblo) PH-13. The preliminary hazards analysis for the WHEAT engineering design package was prepared using an appropriate methodology, and means for mitigating hazards are being tracked.

Finding (Pueblo) PH-14. The Parsons/Honeywell WHEAT technology appears to be capable of demilitarizing the chemical weapons stored at Pueblo Chemical Depot. The combination of hydrolysis of agent, hydrolysis of energetics, and biotreatment of hydrolysates can be a safe and effective process.

Finding (Pueblo) PH-15. While no outstanding waste disposal issues remain, Parsons/Honeywell is exploring alternative waste disposal options to simplify its process by reducing the number of unit operations. Parsons/Honeywell is looking into the off-site disposal of concentrated brine from the bioreactor effluent evaporator. Off-site disposal of other agent-free waste streams, such as hydrolysates, may also be possible, thereby reducing the number or capacity of treatment operations.

Recommendation (Pueblo) PH-1. The residence time in the energetics rotary deactivator (ERD) of the Parsons/Honeywell process must be sufficient for the thermal mass of the fuzes and bursters to rise to the operating temperature, based on the feed rate. The residence time and maximum temperature should be reevaluated and modified to ensure that no energetic materials are present when the metal parts leave the ERD. In addition, feed rate parameters must be adjusted to accommodate overpressure events during thermal deactivation.

Recommendation (Pueblo) PH-2. The detailed design of reactors for the hydrolysis of mustard agent at the Aberdeen, Maryland, storage site should be fully considered in the design of the agent hydrolysis process. A careful evaluation of construction materials should be done.

Recommendation (Pueblo) PH-3. The energetics hydrolysis studies at Holston Army Ammunition Plant should be taken into consideration in the design and preparation of operating procedures for the Parsons/Honeywell energetics hydrolysis system.

Recommendation (Pueblo) PH-4. The feed chutes to the shredders used in the Parsons/Honeywell process must be designed to avoid jamming. Tests should be run with chutes, shredders, and feed materials the size of those in the full-scale operations to ensure that the chute design is adequate.

Recommendation (Pueblo) PH-5. Tests should be carried out to determine that Hastelloy C-276 will adequately resist corrosion in the continuous steam treater.

Recommendation (Pueblo) PH-6. The amount of activated carbon being converted to flammable gas in the continuous steam treater should be measured and its impact on the process assessed.

Recommendation (Pueblo) PH-7. Results of the completed tests of engineering design studies should be carefully reviewed to ensure that activated carbon adsorbers do not have to be added to the bioreactor catalytic oxidation units to achieve acceptable levels of dioxins and furans.

Recommendation (Pueblo) PH-8. A quantitative risk assessment should be developed as early as possible for the operation of the propellant removal room to identify contributors to worker risk and provide guidance for reducing risks by means of design.

5

General Findings and Recommendations

In this chapter, the committee provides general findings on the two technology packages that have undergone EDS testing. The committee also reassesses the findings and recommendations in the two ACW I Committee reports. The general findings below must be considered with acknowledgment of the fact that some ACWA EDS testing was not completed in time for the committee to obtain final test results and that some process steps remain to be demonstrated on a pilot scale.

The energetics hydrolysis test program is progressing at a pace satisfactory to meet the engineering requirements for construction of a disposal facility at Pueblo Chemical Depot. Issues concerning hydrolysis of neat tetryl, optimum granulation sizes, more complete characterization of hydrolysis products from aromatic nitro compounds, and optimum process control strategies for full-scale operations are yet to be investigated.

ENGINEERING DESIGN STUDIES

General Finding (Pueblo) 1. Based on the results of the demonstration tests, the engineering design package, and available data, the committee believes that the Parsons/Honeywell WHEAT technology package can provide an effective and safe means of destruction for the assembled chemical weapons stored at the Pueblo Chemical Depot. However, some of the process steps remain to be demonstrated.

The Parsons/Honeywell technology process provides effective means to:

- disassemble munitions by a modified baseline disassembly process that removes the agent from the projectile bodies by washout
- destroy chemical agent HD to a 99.9999 percent DRE by hydrolysis
- destroy fuzes with the energetics rotary deactivator
- destroy energetic materials to a 99.999 percent DRE by hydrolysis in 15 weight percent hot caustic solution, provided that the following safeguards are observed:
  — different energetic materials are not processed together
  — precautions are taken to ensure that all emulsified TNT is completely destroyed
- control the very large volumes of off-gases emitted from the biotreatment plant through a CATOX unit

However, the committee notes that the effectiveness of some process steps, including removal of energetics from munitions, has not been tested during the EDS. Treatment of metal parts, dunnage, and DPE suit material remains to be demonstrated. No tests are currently planned to demonstrate the efficacy of the burster washout and energetic materials size-reduction steps. The projectile washout system is currently being tested. Other remaining munition disassembly operations are very similar to those used in the baseline system and have therefore been proven. The energetics rotary deactivator concept appears workable but has not been demonstrated at the pilot scale. Energetics hydrolysis is relatively immature, but current testing at Holston AAP has the capability to resolve many, but not all, of these issues (see Chapter 2).

The testing of the continuous steam treater for dunnage and the projectile washout system will not be complete until October 2001. Dioxins and furans are present in the off-gas from the CATOX units on the bioreactors but are below levels of regulatory concern. The batch metal parts treater for small metal parts is being tested, and preliminary data are encouraging. The carousel fixture for the rotary metal parts treater for large metal parts has not been demonstrated. The use of catalytic oxidizers for various streams is currently being tested, but sufficient test data have not been provided to the committee. Because the honeycomb structure of the CATOX unit is susceptible to plugging, proper design must be employed to prevent particulates from entering the catalyst structure.

General Finding (Pueblo) 2. Based on the results of the demonstration tests, the engineering design package, and available data, the committee believes that many aspects of the General Atomics technology package can be effective and safe for the destruction of assembled chemical weapons at the Pueblo Chemical Depot. However, to achieve prolonged operability of the supercritical water oxidation (SCWO) system as designed will require extensive maintenance. In addition, the SCWO processing of dunnage slurried in energetics hydrolysate, which constitutes the vast majority of the feedstock to be processed, remains unproven. The viability of the General Atomics technology package will depend on acceptable operability of the SCWO systems.

The General Atomics technology process provides effective means to:

- disassemble munitions by a modified baseline disassembly that removes the agent from the projectile bodies by cryofracture.
- destroy chemical agent HD to a 99.9999 percent DRE by hydrolysis
- destroy fuzes with the energetics rotary hydrolyzer
- destroy energetic materials to a 99.999 percent DRE by hydrolysis in 15 weight percent hot caustic solution, provided that the following safeguards are observed:
  —different energetic materials are not processed together
  —precautions are taken to ensure that all emulsified TNT is completely destroyed
- provide effective 5X-level decontamination for munition bodies through the use of an electrically heated discharge conveyor
- readily control the very low volumes of off-gases produced through activated carbon adsorption systems

For dunnage, the materials are shredded and reduced in size to 1.0 mm. The slurry is then fed into the SCWO reactors to destroy all the dunnage.

However, the committee has serious concerns about the SCWO system that is used to process the hydrolysates and the slurried dunnage. At the time this report was prepared, not all of the long-term processing tests had been completed. On the basis of results to date, the committee has concerns about the ability of the SCWO reactor to operate continuously for adequate lengths of time. An additional concern is the ability of the size-reduction system to remove 100 percent of the tramp metal that comes with the dunnage. If the tramp metal is not removed from the dunnage, the committee believes it will clog the injectors of the SCWO system and further reduce the system's online availability.

The SCWO tests that have been performed to date, especially those involving chlorinated organic compounds such as HD hydrolysate, have consistently encountered severe corrosion of the reactor material or plugging of the reactor with salts. General Atomics proposes to solve the problem of plugging by periodically (every 22 hours of operation) reducing the pressure of the reactor to slightly below the critical point of water and flushing with clean water for 2 hours to remove the accumulated salts. The technology provider proposes to deal with the corrosion problem by inserting into the SCWO reactor a sacrificial titanium liner and shutting down at approximately every 140 hours of operation to open the reactor and replace or reverse the liner.[1] In the committee's opinion, the flushing step does not pose an unreasonable operating requirement; however, it considers the need for a liner replacement at six-day intervals to be excessively disruptive and not in keeping with sound principles of effective operation. In the full-scale system, liner replacement will require the following steps:

1. Cooling down and depressurizing the reactor,
2. Unbolting and removing an approximately 16-inch-diameter, several-inch-thick pressure head from the top of the reactor,
3. Withdrawing the 12.5-inch-diameter, 19-foot-long titanium liner from the tubular SCWO reactor,
4. Reinserting the same liner reversed end to end or a new liner,
5. Setting the pressure gasket back into place and reattaching the gasket coolant lines,
6. Resetting and bolting the pressure head onto the reactor,
7. Pressure testing the SCWO reactor to assure proper head seating and sealing, and
8. Restarting the heat-up of the system and restarting the waste feed.

This appears to the committee to be a very time-consuming procedure. The experience of a number of committee members has been that large pieces of high-pressure equipment are very difficult and time consuming to seal. Tests have only been conducted with reactors 2 to 4 inches in diameter. The time required for this procedure at the far larger size of the full-scale SCWO unit is highly uncertain.

General Atomics proposes to build duplicate SCWO reactors so that one is operating while the second is being serviced; however, the committee has reservations about whether this level of redundancy is adequate to maintain the proposed operating schedule.

General Finding (Pueblo) 3. As the ACW I Committee observed, the unit operations in both the General Atomics GATS and the Parsons/Honeywell WHEAT technology packages have never been operated as total integrated

---

[1]The corrosion is restricted to the top part of the liner so each liner can be used twice by opening the reactor and reinstalling it in the reactor with the uncorroded lower part up.

processes. As a consequence, a prolonged period of systemization will be necessary for both to resolve integration issues as they arise, even for apparently straightforward unit operations.

This finding continues to be valid following development of and testing for the EDS design packages for the General Atomics and Parsons/Honeywell technologies. Also, in both cases, some of the routine unit operations have not yet been designed or tested. Thus, although they appear straightforward, these unit operations could require some redesign during systemization.

General Finding (Pueblo) 4. Several of the unit operations in both the General Atomics and Parsons/Honeywell processes are intended to treat process streams that are not unique to the chemical weapons stockpile and that could potentially be treated at existing off-site facilities. These streams include agent-free energetics, dunnage, brines from water recovery, and hydrolysates. Off-site treatment would simplify the overall processes and facilitate process integration by eliminating the need for further development of these unit operations. It might also simplify design requirements to meet safety concerns.

All of the process streams that could potentially be treated off-site have compositions similar to waste streams routinely treated by commercial industrial waste treatment facilities and do not exhibit any unique toxicity. Thus, they could be transported by standard commercial conveyance to commercial facilities that are appropriately permitted to receive the waste.

UPDATE ON GENERAL FINDINGS AND RECOMMENDATIONS OF THE ACW I COMMITTEE

The committee reviewed all of the general findings and recommendations from the ACW I Committee reports for continued applicability and disposition (NRC, 2000). The ACW II Committee's assessment of the status of these prior findings and recommendations is summarized below.

General Findings from the 1999 Initial ACW I Committee Report

General Finding 1. The chemistries of all four of the primary technologies (hydrolysis, SILVER II, plasma arc, and SET), as proposed, can decompose the chemical agents with destruction efficiencies of 99.9999 percent. However, each technology package raises other technical issues that must be resolved. One of the crucial issues is the identity and disposition of by-products [from the chemical agents].

The Demonstration I testing and EDS tests have shown that hydrolysis has achieved 99.9999 percent destruction and removal efficiency of agents.

General Finding 2. The technology base for the hydrolysis of energetic materials is not as mature as it is for chemical agents. Chemical methods of destroying energetics have only been considered recently. Therefore, there has been relatively little experience with the alkaline decomposition of ACWA-specific energetic materials (compared to experience with chemical agents). The following significant issues should be resolved to reduce uncertainties about the effectiveness and safety of using hydrolysis operations for destroying energetic materials:

- the particle size reduction of energetics that must be achieved for proper operation
- the solubility of energetics in specific alkaline solutions
- process design of the unit operation and the identification of processing parameters (such as the degree of agitation and reactor residence time) necessary for complete hydrolysis
- the characterization of actual products and by-products of hydrolysis as a function of the extent of reaction
- the selection of chemical sensors and process control strategies to ensure that the unit operation following hydrolysis can accept the products of hydrolysis
- development of a preventative maintenance program that minimizes the possibility of incidents during the cleanup of accumulated precipitates

The PMACWA has undertaken an extensive technology program in support of this finding. The successful completion of the EDS test program on energetics hydrolysis will provide the data called for in this finding.

General Finding 3. The conditions under which aromatic nitro compounds, such as trinitrotoluene (TNT) or picric acid, will emulsify in the aqueous phase and not be completely hydrolyzed are not well understood. Therefore, this type of material could be present in the output stream from an energetic hydrolysis step.

Precautions must be taken to ensure that all emulsified TNT is destroyed in the reactors.

General Finding 4. The products of hydrolysis of some energetic materials have not been characterized well enough to support simultaneous hydrolysis of different kinds of energetic materials in the same batch reactor.

To be conservative, different energetic materials should not be processed together, particularly if they contain lead compounds, until these concerns can be addressed with experimental data.

General Finding 5. The primary chemical decomposition process in all of the technology packages [is hydrolysis, which] produces environmentally unacceptable reaction products. Therefore, all of the packages are complicated processes that include subsequent treatment step(s) to modify these products.

Secondary treatments are required for both technology packages to produce environmentally acceptable products.

General Finding 6. The waste streams of all of the ACWA technology packages could contain very small amounts of hazardous substances (besides any residual chemical agent). These substances were not fully characterized at the time of this report; therefore, all waste streams must be characterized to ensure that human health and the environment are protected. If more than one phase (gas, liquid, or solid) is present in a waste stream, each phase should be characterized separately.

The Demonstration I tests and the EDS studies have substantially characterized the process streams and waste streams. Characterization should continue throughout the development and systemization process.

General Finding 7. None of the proposed technology packages complies completely with the hold-test-release concept for all gaseous effluents (both process and ventilation effluents).

This finding is still valid. However, as discussed elsewhere in this report and in earlier reports, hold-test-release is not necessary.

General Finding 8. Hold-test-release of gaseous effluents may not ensure against a release of agent or other hazardous material to the atmosphere. No evidence shows that hold-test-release provides a higher level of safety than current continuous monitoring methods for gaseous streams with low levels of contamination. Furthermore, none of the technologies provides for hold-test-release of effluents from ventilation systems that handle large volumes of gases from contaminated process areas.

This finding remains valid.

General Finding 9. Solid salts will be hazardous waste, either because they are derived from hazardous waste (see Chapter 2) or because they leach heavy metals above the levels allowed by the Resource Conservation and Recovery Act Toxicity Characteristic Leaching Procedure. Stabilization—mixing waste with a reagent or reagents to reduce the leachability of heavy metals—will probably be required before the salts can be sent to a landfill. The potentially high chloride and nitrate content of these salts will make the waste difficult to stabilize, and treatability studies will be necessary to determine a proper stabilization formula.

The concentrations of RCRA-regulated heavy metals in the evaporator brine have been found to be very low or nondetectable. This stream, and possibly the salts produced from it, will not be considered hazardous by characteristic.

General Finding 10. Testing, verification, and integration beyond the 1999 demonstration phase will be necessary because the scale-up of a process can present many unexpected challenges, and the ACWA demonstrations were limited in nature.

The issues of integration and scale-up still must be resolved.

General Finding 11. Although a comprehensive quantitative risk assessment (QRA), health risk assessment (HRA), and ecological risk assessment (similar to assessments performed for the baseline process) cannot be completed at this stage of process development, these assessments will have to be performed and refined as process development continues.

This finding is still valid.

General Finding 12. The "optimum" system for a particular chemical weapons storage depot might include a combination of unit operations from the technology packages considered in this report.

This finding is still valid. However, the PMACWA is committed to selecting one of the two technology packages discussed in this report. The ACW II Committee has only been asked to evaluate existing technology packages.

General Finding 13. Some of the ACWA technology providers propose that some effluent streams be used commercially. New or modified regulations may have to be developed to determine if these effluent streams can be recovered or reused.

This finding is not applicable to the processes evaluated in this report.

General Finding 14. An extraordinary commitment of resources will be necessary to complete the destruction of the assembled chemical weapons stockpile in time to meet the current [CWC] deadline using any of the ACWA technology packages. This would demand a concerted national effort. It is unlikely that any of the technology packages could meet this deadline.

This finding is still valid.

General Finding 15. The Dialogue process for identifying an alternative technology is likely to reduce the level of public opposition to that technology. The committee believes that the Dialogue has been and continues to be a positive force for public acceptance of alternatives to incineration. Although the Dialogue process requires a significant commitment of time and resources, it has been a critical component of the ACWA program to date.

The Dialogue process for public involvement has been instrumental in accelerating the development of non-incineration processes. Some form of public involvement should be continued throughout the construction phase and operation of the Pueblo facility.

General Finding 16. Although the committee did not have access to scientific data on the attributes of a technology that would be most acceptable to the public, input from members of the active publics and previous research indicates that technologies with the following characteristics are likely to stimulate less public opposition:

- minimal emissions, particularly gaseous
- continuous monitoring of effluents to verify that the process is operating as designed (process assurance measurement)
- provisions for representatives of the local community to observe and participate in the process assurance measurement

This finding remains true. The two processes under consideration have followed these guidelines.

General Recommendations from the 1999 Initial ACW I Committee Report

General Recommendation 1. If a decision is made to move forward with any of the ACWA technology packages, substantial additional testing, verification, and especially integration should be performed prior to full-scale implementation (see General Finding 10).

EDS has provided additional testing. However, not all unit operations have been tested, and system integration has not yet been demonstrated.

General Recommendation 2. The sampling and analysis programs at each phase of development should be carefully reviewed to ensure that the characterization of trace components is as comprehensive as possible to avoid surprises in the implementation of the selected technology (see General Finding 6).

Extensive sampling and analysis have been performed during both Demonstration I and the EDS testing. Characterization of all streams should continue during development and systemization of the processes.

General Recommendation 3. If a decision is made to move forward with any of these technology packages, health and safety evaluations should progress from qualitative assessments to more quantitative assessments as the process design matures. Quantitative (QRA), health (HRA), and ecological risk assessments should be conducted as soon as it is practical. Early initiation of these assessments will allow findings to be implemented with minimal cost and schedule impact (see General Finding 11).

Both technology providers have been performing preliminary hazards analyses.

General Recommendation 4. Any of these technology packages, or any component of these technology packages, should be selected on a site-specific basis (see General Finding 12).

PMACWA is in the process of selecting the technology best suited to destroy the stockpile at Pueblo Chemical Depot.

General Recommendation 5. Whatever unit operation immediately follows the hydrolysis of energetic materials should be designed to accept emulsified aromatic nitro compounds, such as TNT or picric acid, as contaminants in the aqueous feed stream (see General Finding 3).

Precautions must be taken to ensure that all emulsified aromatic nitro compounds are destroyed in the reactors.

General Recommendation 6. Simultaneous processing of different types of energetic materials should not be performed until there is substantial evidence that the intermediates formed from the hydrolysis of aromatic nitro compounds will not combine with M28 propellant additives or ordnance fuze components to form extremely sensitive explosives, such as lead picrate (see General Finding 4).

This remains a concern. Different energetic materials should not be processed together.

General Findings from the 2000 Supplemental ACW I Committee Report

General (Demo I) Finding 1. Based on the committee's assessment of the maturity of the various unit operations, none of the three technology packages is ready for *integrated* pilot programming, although certain unit operations are sufficiently mature to bypass pilot testing (e.g., hydrolysis of agent).

This finding has been updated by the findings in this report.

General (Demo I) Finding 2. The demonstration tests were not operated long enough to demonstrate reliability and long-term operation.

This finding is valid.

General (Demo I) Finding 3. The committee reiterates that none of the unit operations have yet been integrated into a complete system. The lack of integration remains a major concern as a significant obstacle to full-scale implementation.

This finding is still valid.

References

Belcher, L. 2000. PMACWA Hydrolysis at Pantex. Briefing by L. Belcher, Department of Energy BWXT Pantex Plant, to the Hydrolysis Review Meeting, Holston Army Ammunition Plant, Kingsport, Tenn., October 31.

Bishop, R.L. 2000. Bench-Scale Base Hydrolysis. Briefing by R.L. Bishop, High Explosive Science and Technology, Los Alamos National Laboratory, to the Committee on Review and Evaluation of Alternative Technologies for Demilitarization of Assembled Chemical Weapons, Phase II, National Research Council, Woods Hole, Mass., October 19.

Bishop, R.L. 2001. Briefing by R.L. Bishop, High Explosive Science and Technology, Los Alamos National Laboratory, to the Committee on Review and Evaluation of Alternative Technologies for Demilitarization of Assembled Chemical Weapons, Phase II, National Research Council, Washington, D.C., February 8.

Bishop, R.L., R.L. Flesner, S.A. Larson, and D.A. Bell. 2000. Base hydrolysis of TNT-based explosives. Journal of Energetic Materials 18(4): 275–288.

Bonnett, P.C. 2000. Test Plan Requirements: Characterization of an Energetic Hydrolysis Reactor System at Holston Army Ammunition Plant, July 12. Picatinny Arsenal, N.J.: U.S. Army TACOM-ARDEC.

Bonnett, P.C. 2001. Energetics Hydrolysis, Tests in Progress. Briefing by P.C. Bonnett, Engineer, U.S. Army TACOM-ARDEC, to the Committee on Review and Evaluation of Alternative Technologies for Demilitarization of Assembled Chemical Weapons, Phase II, National Research Council, Washington, D.C., February 8.

Burns and Roe. 1999. Assembled Chemical Weapons Assessment Program Final Report, June. Oradell, N.J.: Burns and Roe Enterprises, Inc.

Crooker, P.J., K.S. Ahluwalia, and Z. Fan. 2000. Final results of shipboard hydrothermal oxidation with a transpiring wall reactor. Paper # 15A-2 in Proceedings of the Conference on Incineration and Thermal Treatment Technologies (IT-3), Portland, Ore., May 8–12. Irvine, Calif.: University of California, Department of Environmental Health and Safety.

Dobratz, B.M. 1981. Lawrence Livermore National Laboratory Explosives Handbook: Properties of Chemical Explosives and Explosive Simulants. Livermore, Calif.: Lawrence Livermore National Laboratory.

DOD (Department of Defense). 1993. Military Standard 882C: Standard Practice for System Safety Program Requirements, January. Washington, D.C.: DOD.

DOD. 1997. Assembled Chemical Weapons Assessment Program Annual Report to Congress, December. Aberdeen Proving Ground, Md.: Program Manager for Assembled Chemical Weapons Assessment.

DOD. 1998. Assembled Chemical Weapons Assessment Program Annual Report to Congress, December. Aberdeen Proving Ground, Md.: Program Manager for Assembled Chemical Weapons Assessment.

DOD. 1999. Assembled Chemical Weapons Assessment Program Supplemental Report to Congress, September. Aberdeen Proving Ground, Md.: Program Manager for Assembled Chemical Weapons Assessment.

DOD. 2000. Assembled Chemical Weapons Assessment Program Annual Report to Congress, December. Aberdeen Proving Ground, Md.: Program Manager for Assembled Chemical Weapons Assessment.

Elliott, J.P., D.A. Hazlebeck, D.W. Ordway, A.J. Roberts, and M.H. Spritzer. 2000. Update on hydrothermal oxidation developments on DARPA/ONR and Air Force projects at General Atomics. Paper # 15A-1 in Proceedings of the Conference on Incineration and Thermal Treatment Technologies (IT-3), Portland, Ore., May 8–12. Irvine, Calif.: University of California, Department of Environmental Health and Safety.

EPA (Environmental Protection Agency). 1998. Human Health Risk Assessment Protocol for Hazardous Waste Combustor Facilities. EPA 530-D-98-001B, July. Washington, D.C.: EPA.

Garrison, M.M. 1994. Conventional weapons utilization in Ukraine and Belarus: Recovery-reuse and conversion. Proceedings of the 2nd Demilitarization Symposium, Arlington, Va., May 23-25. Arlington, Va.: National Defense Industrial Association.

General Atomics. 1997. Supercritical Water Oxidation Corrosion Studies: Technical Report, Vols. 1 and 2. Aberdeen Proving Ground, Md.: Program Manager for Assembled Chemical Weapons Assessment.

General Atomics. 1999a. Assembled Chemical Weapons Assessment (ACWA) Draft Test Technical Report, June 30. Aberdeen Proving Ground, Md.: Program Manager for Assembled Chemical Weapons Assessment.

General Atomics. 1999b. Responses to PMACWA Questions and Clarification Request: General Atomics Final Technical Report. Aberdeen Proving Ground, Md.: Program Manager for Assembled Chemical Weapons Assessment.

General Atomics. 2000a. ACWA Engineering Design Study, Final GAS Full-Scale Preliminary Design Engineering Package, October 27. San Diego, Calif.: General Atomics, Inc.

General Atomics. 2000b. Assembled Chemical Weapons Assessment (ACWA), Engineering Design Study Plan, May 25. San Diego, Calif.: General Atomics, Inc.

General Atomics. 2000c. Briefing package, General Atomics Total Solution Design Package Review Meeting, Arthur D. Little, Inc., Cambridge, Mass., August 17–18.

General Atomics. 2000d. Briefing package presented to the Committee on Review and Evaluation of Alternative Technologies for Demilitarization of Assembled Chemical Weapons, Phase II. Parsons Corporation, Pasadena, Calif., October 9–10.

General Atomics. 2000e. Briefing package presented to the Supercritical Water Oxidation In-Process Review Meeting, Program Manager for

REFERENCES

Assembled Chemical Weapons Assessment, Aberdeen Proving Ground, Md., September 6.

General Atomics. 2000f. Briefing package presented to the Supercritical Water Oxidation In-Process Review Meeting, Program Manager for Assembled Chemical Weapons Assessment, Aberdeen Proving Ground, Md., December 7.

General Atomics. 2000g. Briefing package presented to the EDS Design-Review Meeting, Parsons Corporation, Pasadena, Calif., November 8–10.

General Atomics. 2000h. Final GATS Preliminary Hazards Analysis, October. San Diego, Calif.: General Atomics, Inc.

General Atomics. 2001. Assembled Chemical Weapons Assessment (ACWA), Engineering Design Studies, Draft Test Report, February 10. San Diego, Calif.: General Atomics, Inc.

Gibbs, T.R., and A. Popolato. 1980. LASL Explosive Property Data. Berkeley, Calif.: University of California Press.

Goldstein, R. 1999. An Overview of the Army's Demilitarization Technology Research and Development Program for Conventional Ammunition, Proceedings of 5th Life Cycles of Energetic Materials Conference, Orlando, Fla., September 26–29.

Griffith, J.W. 2000. The use of hydrothermal oxidation in the treatment of municipal wastewater sludge. Paper # 15A-3 in Proceedings of the Conference on Incineration and Thermal Treatment Technologies (IT-3), Portland, Ore., May 8–12. Irvine, Calif.: University of California, Department of Environmental Health and Safety.

Hong, G. 2001. ACWA EDS SCWO Testing Progress Update. Briefing by General Atomics, Inc., to the Committee on Review and Evaluation of Alternative Technologies for Demilitarization of Assembled Chemical Weapons, Phase II, National Research Council, Washington, D.C., February 8.

JOCG (Joint Ordnance Commanders Group). 2000. Optimization of DOD Open Burning/Open Detonation Units: Phase II, Analysis and Closure Recommendations, March. Rock Island, Ill.: U.S. Army Operations Support Command.

Kaye, S.M., and H.L. Herman. 1980. Encyclopedia of Explosives and Related Items. PATR 2700, Vol. 9. Dover, N.J.: U.S. Army Armament Research and Development Command.

Macdonald, D, and L. Kriksunov. 2001. Probing the Chemical and Electrochemical Properties of SCWO Systems, Prepublication copy. University Park, Pa.: Pennsylvania State University Center for Electrochemical Science and Technology.

Machacek, O. 2000. Application of Demilitarized Gun and Rocket Propellant in Commercial Explosives. NATO Science Series. II. Mathematics, Physics, and Chemistry – Vol. 3. O. Machacek, ed. Dordrecht, Netherlands: Kluwer Academic Publishers.

Marinkas, A., C. Hu, and C. Smith. 1998. Use of recycled RDX for production of RDX-based explosives. Proceedings of 4th Life Cycles of Energetic Materials Conference, Fullerton, Calif., March 29–April 1.

McCloskey, G. 2000. JACADS Closure Status. Briefing by Gary McCloskey, site project manager, PMCD, to the Committee on Review and Evaluation of the Army Chemical Stockpile Disposal Program, Anniston Chemical Disposal Facility, Anniston, Ala., March.

Mitchell, W.J. 1998. State of the Science and Research Needs in the Characterization and Minimization of the Emissions for Ordnance Use and Disposal Activities. Available online at <http:// www.enviro-engrs.org/Ordnance.pdf> (March 8, 2001).

Myler, C. 2000. ABCDF Updates and Issues. Briefing by Craig Myler, Bechtel, to the Committee on Review and Evaluation of the Army Chemical Stockpile Disposal Program, National Research Council, Washington, D.C., June 22.

NRC (National Research Council). 1991. Demilitarization of Chemical Weapons by Cryofracture: A Technical Assessment. Washington, D.C.: National Academy Press.

NRC. 1993. Alternative Technologies for the Destruction of Chemical Agents and Munitions. Washington, D.C.: National Academy Press.

NRC. 1996. Review and Evaluation of Alternative Chemical Disposal Technologies. Washington, D.C.: National Academy Press.

NRC. 1999. Review and Evaluation of Alternative Technologies for Demilitarization of Assembled Chemical Weapons. Washington, D.C.: National Academy Press.

NRC. 2000. Evaluation of Demonstration Test Results of Alternative Technologies for Demilitarization of Assembled Chemical Weapons. Washington, D.C.: National Academy Press.

NRC. 2001. SCWO EST. Letter report to Dr. Hank Dubin, director, Assessments and Evaluation, Office of the Assistant Secretary of the Army for Acquisition, Logistics and Technology (SAAL), January 29. Washington, D.C.: Board on Army Science and Technology.

Newman K.E., K.T. Kenar, S.J. Rosenberg, and Y.B. Yim. 1997. Feasibility of Reclamation and Reuse of RDX for Joint Mine Countermeasure Programs (IHTR 2036), October 31. Indian Head, Md.: Naval Surface Warfare Center.

Newman, K. 1999. A Review of Alkaline Hydrolysis of Energetic Materials: Is It Applicable to Demilitarization of Ordnance? IHTR 2167, April 30. Indian Head, Md.: Naval Surface Warfare Center.

Parsons. 2000a. Briefing package presented to EDS Design Review Meeting, Parsons Corporation, Pasadena, Calif., November 8–9.

Parsons. 2000b. Project Design Note (T)—002, May 18. Pasadena, Calif.: Parsons Infrastructure and Technology Group.

Parsons. 2000c. Process Hazardous Analysis: Process Analysis, October 23. Pasadena, Calif.: Parsons Infrastructure and Technology Group.

Parsons. 2000d. Client Review Submittal, ACWA EDS WHEAT Design Basis, October. Pasadena, Calif.: Parsons Infrastructure and Technology Group.

Parsons. 2000e. Assessment of Technologies for Assembled Chemical Weapon Demilitarization, EDS Test Report: Interim Report 3.2, Immobilized Cell Bioreactor (ICB), December 13. Pasadena, Calif.: Parsons Infrastructure and Technology Group.

Parsons. 2000f. Project Design Note (T)—005, June 28. Pasadena, Calif.: Parsons Infrastructure and Technology Group.

Parsons. 2000g. Assessment of Technologies for Assembled Chemical Weapon Demilitarization, EDS Test Report: Interim Report 3.3, Catalytic Oxidation (CATOX), December 13. Pasadena, Calif.: Parsons Infrastructure and Technology Group.

Parsons-AlliedSignal. 1999. Assessment of Technologies for Assembled Chemical Weapons Demilitarization Demonstration Test. Final Report, July 1. Pasadena, Calif.: Parsons Infrastructure and Technology Group.

PMACWA (Program Manager for Assembled Chemical Weapons Assessment). 2000. SCWO EST Site Daily Update, December 2. Aberdeen Proving Ground, Md.: Program Manager for Assembled Chemical Weapons Assessment.

PMCD (Program Manager for Chemical Demilitarization). 1996. PMCD Chemical Agent Disposal Facility Risk Management Program Requirements, May. Aberdeen Proving Ground, Md.: Program Manager for Chemical Demilitarization.

PMCD. 1997. PMCD Guide to Risk Management Policy and Activities. Rev. 0, May. Aberdeen Proving Ground, Md.: Program Manager for Chemical Demilitarization.

Spritzer, M.H. 2000a. GATS Engineering Design Package. Briefing by M.H. Spritzer, manager, Demilitarization Technology, General Atomics, Inc., to the General Atomics Total Solution Design Package Review Meeting, Arthur D. Little, Inc., Cambridge, Mass., August 14–15, 2000.

Spritzer, M.H. 2000b. Responses to questions on General Atomics design. Briefing by M.H. Spritzer, manager, Demilitarization Technology, General Atomics, Inc., to the Committee on Review and Evaluation of Alternative Technologies for Demilitarization of Assembled Chemical Weapons, Phase II, Parsons Corporation, Pasadena, Calif., November 26–28, 2000.

Susman, S. 2000a. Engineering Design Study. Briefing by S. Susman, mechanical engineer, ACWA Technical Team, Program Manager for

Assembled Chemical Weapons Assessment, to the Committee on Review and Evaluation of Alternative Technologies for Demilitarization of Assembled Chemical Weapons, Phase II, National Research Council, Washington, D.C., August 9.

Susman, S. 2000b. Engineering Design Study Status. Briefing by S. Susman, mechanical engineer, ACWA Technical Team, Program Manager for Assembled Chemical Weapons Assessment, to the Committee on Review and Evaluation of Alternative Technologies for Demilitarization of Assembled Chemical Weapons, Phase II, National Research Council, Woods Hole, Mass., October 19.

SWEC (Stone & Webster Engineering Corporation). 1996. Supercritical Water Oxidation Data Acquisition Testing, Final Report. Vol. 1, Phase I DOE Testing for U.S. DOE Idaho Operations Office and Strategic Environmental Research and Development Program. Baton Rouge, La.: Stone & Webster Engineering Corporation.

U.S. Army. 1988. Chemical Stockpile Disposal Program Final Programmatic Environmental Impact Statement. Aberdeen Proving Ground, Md.: Program Manager for Chemical Demilitarization.

U.S. Army. 1997. Assessment of Technologies for Assembled Chemical Weapons Demilitarization. Solicitation Number DAAM01-97-R-0031, July 28. Aberdeen Proving Ground, Md.: U.S. Army Chemical and Biological Defense Command.

U.S. Army. 1998. Fact Sheet: Tooele Chemical Agent Disposal Facility Byproducts and Waste Streams. Aberdeen Proving Ground, Md.: Program Manager for Chemical Demilitarization.

Wojciechowski, P. 2000. NECDF Updates and Issues. Briefing by Paul Wojciechowski, Deputy NECDF Manager, Parsons Infrastructure and Technology Group, Inc., to the Committee on Review and Evaluation of the Army Chemical Stockpile Disposal Program, National Research Council, Washington, D.C., June 22.

Appendixes

Appendix A

Description of Munitions in the Pueblo Chemical Depot Stockpile

Figures A-1 through A-3 are cutaway drawings of the 105-mm shell, 155-mm shell, and 4.2-inch mortar projectiles. Information is also included on the size, weight, energetics, and packaging of each projectile.

The stockpile inventory at Pueblo Chemical Depot consists entirely of munitions containing mustard agent. Most of the projectiles contain mustard agent HD (distilled β,β'-dichloroethyl sulfide). Some contain mustard agent HT, a 60:40 eutectic mixture of HD and bis(2-[2-chloroethylthio]ethyl)ether. All of the munitions may contain some degradation products and inorganic residues.

REFERENCE

U.S. Army. 1977. Army Ammunition Data Sheets: Artillery Ammunition, Guns, Howitzers, Mortars, Recoilless Rifles, Grenade Launchers, and Artillery Fuzes (FSC 1310, 1315, 1320, 1390). TM 43-0001-28. April 1977. Washington, D.C.: Headquarters, U.S. Army.

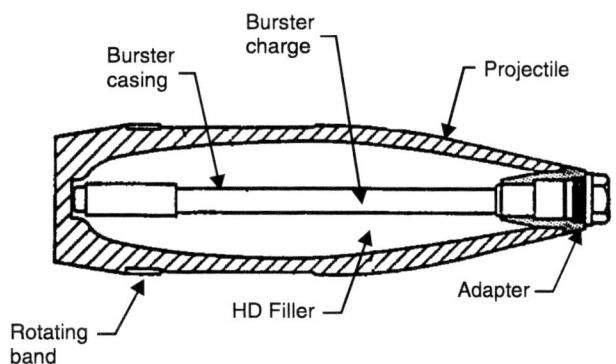

M60 Cartridge, 105-mm Howitzer

Length	31.1 inches	Booster	M22
Diameter	105 mm	Explosive	Tetrytol
Total weight	42.92 lb	Explosive weight	0.3 lb
Agent	HD	Propellant	M67
Agent weight	2.97 lb	Propellant weight	2.83 lb
Fuze	M557/M51A5	Primer	M28A2/M28B2
Burster	M5	Packaging	1 round/fiber container, 2 container/wooden box

FIGURE A-1 105-mm howitzer projectile. Note: M60 105-mm cartridges have been reconfigured and therefore will not have propellant attached. Source: Adapted from U.S. Army, 1977.

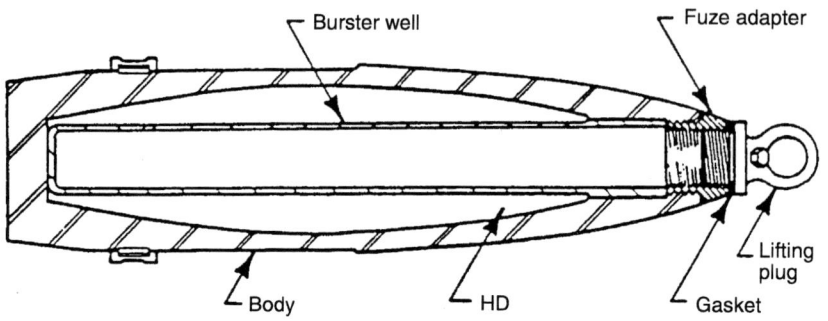

M110 Projectile, 155-mm Howitzer

Length	31.1 inches	Booster	M22
Diameter	155 mm	Explosive weight	0.41 lb
Total weight	94.6 lb	Propellant	None
Agent	HD	Propellant weight	None
Agent weight	11.7 lb	Primer	None
Fuze	None	Packaging	8 rounds/wooden pallet
Burster	M6		

FIGURE A-2 155-mm howitzer projectile. Source: Adapted from U.S. Army, 1977.

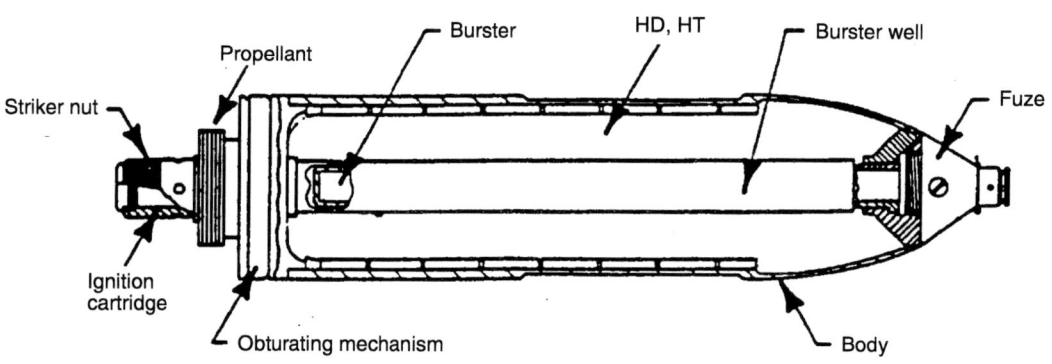

Cartridge, 4.2-inch Cartridge/Mortar

	M2/HT	M2A1/HD
Length	21.0 inches	21.0 inches
Diameter	4.2 inches	4.2 inches
Total weight	24.67 lb	24.67 lb
Agent	HT	HD
Agent weight	5.8 lb	6.0 lb
Fuze	M8	M8
Burster	M14	M14
Explosive	Tetryl	Tetryl
Explosive weight	0.14 lb	0.14 lb
Propellant	M6	M6
Propellant weight	0.6 lb	0.4 lb
Primer	M2	M2
Packaging	1 round/fiber container, 2 containers/wooden box	1 round/fiber container, 2 containers/wooden box

FIGURE A-3 4.2-inch mortar cartridge. Note: 4.2-inch cartridges/mortars will be reconfigured as projectiles. Most 4.2-inch cartridges will also be defuzed. Source: Adapted from U.S. Army, 1977.

Appendix B

SCWO Reliability and Maintenance (RAM) Log for 500-Hour HD Hydrolysate Run

The data in Table B-1 were collected during the General Atomics engineering design study testing on the use of supercritical water oxidation (SCWO) technology as a secondary treatment step in the General Atomics Total Solution (GATS) for disposal of the mustard-agent-containing munitions stored at the Pueblo Chemical Depot.

REFERENCE

General Atomics. 2001. Assembled Chemical Weapons Assessment (ACWA), Engineering Design Studies, Draft Test Report, February 10. San Diego, Calif: General Atomics, Inc.

TABLE B-1 SCWO Reliability and Maintenance (RAM) Log for 500-Hour HD Hydrolysate Run

ID	Date	Time	Description of Item	GA Log Book No.	Page No.	Shutdown Caused?	Fix Time (hr)	Problem Type	System/ Component Involved
1	1/4/01	0:56	Nozzle plug	11574	56	Y	<1	Operational	Feed nozzle
2	1/5/01	12:07	Liquid letdown valve (Badger) repair	11574	58	N	2	Mechanical	Liquid letdown
3	1/5/01	12:07	Liquid letdown valve cannot hold level	11574	58	N	2	Mechanical	Liquid letdown
4	1/5/01	14:30	Gas/liquid separator gasket fails	11574	58	N	4	Mechanical	Gas/liquid separator
5	1/5/01	17:00	Low-pressure feed pump housing cracked	NA	NA	N	2	Error	Feed recirculation pump
6	1/5/01	17:40	Feed pump erratic	11574	59	N	<1	Operational	Feed pump
7	1/6/01	4:00	Gas analyzers erratic	11574	62	N	<1	Operational	Gas analysis train
8	1/6/01	8:10	TE-600 faulty reading	11574	62	N	<1	Operational	Reactor
9	1/6/01	8:20	Heat tracing on hot line	11574	62	N	<1	Error	Reactor
10	1/6/01	8:20	PT-463 faulty reading	11574	62	N	<1	Electrical	Air line
11	1/6/01	9:15	Letdown system solids accumulation	11574	63	N	1	Operational	GLS, hydroclone
12	1/6/01	9:18	Fuel pump packing repair	11574	63	N	2	Mechanical	Fuel pump
13	1/6/01	10:18	Gas analyzers not consistent with computer	11574	64	N	<1	Electrical	Gas analysis train
14	1/6/01	10:55	Feed pump erratic	11574	64	N	<1	Operational	Feed pump
15	1/6/01	16:38	Liquid letdown valve (Badger) repair	11574	66	N	2	Mechanical	Liquid letdown
16	1/6/01	18:45	Fuel line plug	11574	66	Y	2	Operational	Fuel tank
17	1/6/01	22:00	Fuel pump packing repair	NA	NA	N	2	Mechanical	Fuel pump
18	1/6/01	23:30	Gas sample line plug	11575	2	N	<1	Operational	Gas analysis train
19	1/7/01	14:20	Reactor plug	11574	71	Y	3	Operational	Reactor
20	1/7/01	19:23	Air flow control poor	11574	71	N	<1	Operational	Air line
21	1/8/01	12:49	Reactor bypassing	11574	75	Y	4	Operational	Reactor
22	1/8/01	15:33	Thermowell replacement	11574	76	N	<1	Corrosion	Reactor
23	1/8/01	18:00	Letdown system solids accumulation	NA	NA	N	1	Operational	GLS, hydroclone
24	1/8/01	20:33	Feed cross leak	11574	76	N	2	Error	Reactor
25	1/8/01	21:57	Reactor dump valve leak	11574	76	N	<1	Mechanical	Reactor
26	1/8/01	22:49	Nozzle plug	11574	76	N	4	Operational	Feed nozzle
27	1/8/01	22:57	Water pump ruptured diaphragm	11574	76	N	2	Mechanical	Water pump
28	1/9/01	2:32	Reactor surface TEs (thermocouples) faulty readings	11574	77	N	1	Electrical	Reactor
29	1/9/01	3:27	Badgers not working in auto mode	11574	77	N	<1	Computer	Computer
30	1/9/01	4:45	Computer type mismatch	11574	77	N	<1	Computer	Computer
31	1/9/01	6:47	Feed pump erratic	11574	77	N	1	Operational	Feed pump
32	1/9/01	6:50	Computer type mismatch	11574	77	Y	<1	Computer	Computer
33	1/9/01	9:50	Low-pressure feed pump seized	11574	78	N	1	Operational	Feed recirculation pump
34	1/10/01	10:00	Gas sample line plug	11574	85	N	<1	Operational	Gas analysis train
35	1/11/01	0:00	pH electrode fouled	NA	NA	N	<1	Operational	Liquid analysis train
36	1/11/01	16:36	Gas sample line plug	11574	94	N	<1	Operational	Gas analysis train
37	1/11/01	5:30	Clean P-350B check valves	11574	98	N	1	Operational	Feed pump
38	1/12/01	7:00	Badgers not working in auto mode	11574	99	N	<1	Computer	Computer
39	1/12/01	7:00	Bypassing behind Ti liner	11574	99	N	<1	Mechanical	Reactor
40	1/12/01	7:00	Potential computer problems	11574	99	N	<1	Computer	Computer
41	1/12/01	7:00	Thermowell replacement	11574	99	N	<1	Corrosion	Reactor
42	1/12/01	10:32	Liner thermocouple malfunction	11574	99	N	1	Mechanical	Reactor
43	1/12/01	15:07	pH meter apparently broken	11574	100	N	2	Mechanical	pH meter
44	1/12/01	15:35	Fuel tank runs dry	11574	100	Y	3	Error	Fuel system
45	1/12/01	19:00	PCV-437A not holding pressure	NA	NA	N	<1	Operational	Fuel system
46	1/12/01	19:39	Ruptured cooling water hose	11574	101	Y	<1	Operational	Cooldown HX
47	1/12/01	22:00	Added scaling chemicals to cooling water	NA	NA	N	<1	Operational	Cooling water system
48	1/13/01	14:14	Gas sample line plug	11575	2	N	<1	Operational	Gas analysis train
49	1/13/01	14:49	Compressor filters need replacing	NA	2	N	1	Operational	HP air compressors
50	1/13/01	21:58	PCV-357B not holding pressure	11575	4	N	<1	Operational	Feed system
51	1/13/01	21:58	Clean P-350B check valves	11575	4	N	1	Operational	Feed pump
52	1/13/01	21:58	P-420B B head leaking	11575	4	N	<1	Operational	Fuel pump
53	1/14/01	18:30	Air flow fluctuations	11575	10	N	<1	Operational	HP air compressors
54	1/14/01	23:50	Discharge and suction check valves cleaned	NA	NA	N	<1	Operational	Feed system
55	1/15/01	4:00	Added scaling chemicals to cooling water	NA	NA	N	<1	Operational	Cooling water system

Component ID	Maintenance/ Fix Type	Root Cause of Problem	Resolution to Problem
NA	Operational modification	Feed precipitation	Leave preheaters on at low setting
FCV-654A	Change-out	Faulty installation, stem bent	Change-out valve, rebuild correctly
FCV-654B	Change-out trim	Solids in valve	Better solids separation prior to valve
GLS-630	Replace gasket	Temporary gasket in use	Procure proper gasket
P-300B	Replace	Pump overpressured during cleanout	Replace pump, exercise care when blowing out lines
P-350B	Clean (flush)	Solids in pump	Periodic cleaning/flushing, including inlet strainer
GS-680	Drain	Water carryover	Modify post-V-680 trap
TE-600	Reposition	Thermocouple out of position	Replace and secure thermocouple
NA	Reposition	Incorrect installation	Reposition heat tracing
PT-463	Replace	TBD	TBD
GLS-630, SS-640	Clean (flush)	Titanium corrosion	Periodic flushing
P-420B	Repair	Normal wear	Normal maintenance
GS-680	Check connections	Connections incorrect	Correct connections
P-350B	Clean (flush)	Solids in pump	Periodic cleaning/flushing, including inlet strainer
FCV-654A	Adjustment	Improper adjustment	Adjust
TK-400B	Clean	Dirty fuel	Screen fuel when adding
P-420A	Repair	Normal wear	Normal maintenance
V-680	System modification	Mist carryover with salt deposition	Modify post-V-680 trap
R-550	Clean (flush)	Failure to rinse after prior run	Rinse reactor before restarting if feed processed for >0.5 hr
FCV-461	Adjustment	Improper adjustment of control pressure	Reset control pressure
R-550	Clean (flush)	Nonoptimal top insert and gasket	Install new top insert and gasket
R-550	Replace	Corrosion	Replace
GLS-630, SS-640	Clean (flush)	Titanium corrosion	Periodic flushing
R-550	Replace	Improper installation	Review assembly procedures with pipe fitters
FV-550	Replace	Normal wear	Rebuild
NA	Clean (flush)	Metal burr from feed cross (see no. 15)	Assemble fittings without galling
P-120B	Repair	Nozzle plug	Assemble fittings without galling
TE-554	Replace	Wet leads	Waterproof thermocouple connections
NA	Programming	Software problem	Edit program
NA	Programming	Software problem	Switch computers
P-350B	Clean (flush)	Solids in pump	Periodic cleaning/flushing, including inlet strainer
NA	Programming	Software problem	Edit program
P-300B	Clean (flush)	Solids precipitation in pump and lines	Clean
V-680	System modification	Mist carryover with salt deposition	Modify post-V-680 trap
AE-670	Clean (flush)	Solids buildup in probe housing	Periodic cleaning/flushing
V-680	System modification	Mist carryover with salt deposition	Modify post-V-680 trap
P-350B	Clean	Preventative maintenance	Clean
NA	Programming	Software problem	Edit program
R-550	Redesign	Mismatched components	New top insert and gasket
NA	Reboot	Power fluctuations	Reboot
R-550	Replace	Corrosion	Replace
TE-572x, 574x	Replace	Normal wear	Replace
AT-670	Replace	Unknown	Not resolved at this time
TK-400B	Refill	Hourly rounds neglected	Carry out hourly rounds
PCV-437A	Repair	Stem and seat worn	Rebuild
HX-610	Repair	Hose overheats, presumably blockage	Rework hoses
Cooling water	Added Nalco chemicals	Potential scaling in cooling water	Treat cooling water
V-680	System modification	Mist carryover with salt deposition	Modify post-V-680 trap
C-450	Change filter	Preventative maintenance	Changed desiccant air dryer filters on all compressors
PCV-357	Rebuild	Stem and seat worn	Rebuild
P-350B	Clean	Preventative maintenance	Clean
P-420B	Replace packing	B head leaking (packing)	Changed packing on B head
C-450	Adjustment	Insufficient compressor pressure	Set start pressure lower
P-350B	Clean	Preventative maintenance	Disassemble and clean
Cooling water	Added Nalco chemicals	Potential scaling in cooling water	Treat cooling water

continued

TABLE B-1 Continued

ID	Date	Time	Description of Item	GA Log Book No.	Page No.	Shutdown Caused?	Fix Time (hr)	Problem Type	System/Component Involved
56	1/15/01	16:24	Gas sample line plug	11575	17	N	<1	Operational	Gas analysis train
57	1/16/01	0:35	Scaling in the DPT lines	11575	20	N	<1	Operational	Reactor DPT
58	1/16/01	0:35	Hose replacement for cooling water to CDHX	11575	20	N	1	Operational	Cooling water system
59	1/16/01	0:35	Bad trim for FCV-654B	11575	20	N	<1	Operational	Liquid letdown
60	1/16/01	0:35	PCV-357B not holding pressure	11575	20	N	<1	Operational	Feed system
61	1/16/01	0:35	Discharge and suction check valves cleaned	11575	20	N	<1	Operational	Feed system
62	1/16/01	0:35	Thermowell replacement	11575	20	N	<1	Corrosion	Reactor
63	1/16/01	0:35	TE-571 broken on disassembly	11575	20	N	<1	Mechanical	Reactor
64	1/16/01	11:00	PCV-357A not holding pressure	NA	NA	N	<1	Operational	Feed system
65	1/16/01	17:27	Gas sample line plug	11575	26	N	<1	Operational	Gas analysis train
66	1/16/01	3:00	Discharge and suction check valves cleaned	11575	28	N	<1	Operational	Feed system
67	1/17/01	3:00	Added scaling chemicals to cooling water	11575	28	N	<1	Operational	Cooling water system
68	1/17/01	12:17	Gas sample line plug	11575	31	N	<1	Operational	Gas analysis train
69	1/17/01	17:00	HP air compressor C malfunction	11575	32	N	2	Mechanical	HP air compressors
70	1/18/01	2:45	PCV-460 not controlling properly	11575	35	N	<1	Operational	HP air line to reactor
71	1/18/01	9:35	Gas sample line plug	11575	37	N	<1	Operational	Gas analysis train
72	1/19/01	5:00	TE-571 broken on disassembly	11575	42	N	<1	Mechanical	Reactor
73	1/19/01	5:00	Thermowell replacement	11575	42	N	<1	Corrosion	Reactor
74	1/20/01	12:15	pH meter apparently broken	11575	48	N	2	Mechanical	pH meter
75	1/20/01	14:00	TE-641 malfunction	11575	48	N	<1	Mechanical	Letdown system
76	1/20/01	19:27	Gas sample line plug	11575	49	N	<1	Operational	Gas analysis train
77	1/20/01	20:10	Line from PT-551 to reactor head leaking	11575	50	Y	2	Mechanical	Reactor plumbing
78	1/20/01	22:12	Shutoff valve on gas letdown leaking	11575	50	N	2	Mechanical	Gas letdown
79	1/20/01	22:12	Reactor dump valve leaking	11575	50	N	2	Mechanical	Reactor
80	1/20/01	22:12	BPR on gas letdown not holding pressure	11575	50	N	2	Mechanical	Gas letdown

Source: General Atomics, 2001.

APPENDIX B

Component ID	Maintenance/ Fix Type	Root Cause of Problem	Resolution to Problem
V-680	System modification	Mist carryover with salt deposition	Modify post-V-680 trap
DPT-550	Clean (flush)	Salts build up in DPT lines	Flush lines
HX-610	Replace	Possible coolant overheating	Replace with high-temperature hose
FCV-654B	Replace	Scored trim, trim too large	Replace with M trim
PCV-357B	Clean	Dirty stem and seat	Disassemble and clean
P-350B	Clean	Preventative maintenance	Disassemble and clean
R-550	Replace	Corrosion	Replace
TE-571	Replace	Thermowell corrosion	Replace
PCV-357A	Clean	Dirty stem and seat	Disassemble and clean
V-680	System modification	Mist carryover with salt deposition	Modify post-V-680 trap
P-350B	Clean	Preventative maintenance	Disassemble and clean
Cooling water	Added Nalco chemicals	Potential scaling in cooling water	Treat cooling water
V-680	System modification	Mist carryover with salt deposition	Modify post-V-680 trap
C-450C	Repair	Check valve broken	Repair
PCV-460	Replace	Oil contamination/clogging frit	Disassemble and rebuild
V-680	System modification	Mist carryover with salt deposition	Modify post-V-680 trap
TE-571	Replace	Thermowell corrosion	Replace
R-550	Replace	Corrosion	Replace
AT-670	Replace	Interference from power wiring	Rearrange wiring
TE-641	Replace	Unknown	Replace with surface thermocouple
V-680	System modification	Mist carryover with salt deposition	Modify post-V-680 trap
NA	Replace	Fatigue/wear/corrosion	Weld new fitting to modified AE drilled through plug
FV-640B	Replace	Contamination in gas letdown	Replace
FV-550	Replace body	Overheating soft goods	Replace
PCV-646	Replace	Corrosion/erosion internals	Replace

Appendix C

Committee Meetings and Site Visits

Committee Meeting 1, June 8–9, 2000
Edgewood, Maryland

NRC Participants

Committee Chair: Robert A. Beaudet. Committee members: Richard J. Ayen, Joan B. Berkowitz, Sheldon E. Isakoff, David S. Kosson, Frederick J. Krambeck, John A. Merson, William R. Rhyne, William R. Seeker, and Leo Weitzman. NRC staff members: Patricia P. Paulette, Bruce Braun, Harrison T. Pannella, and Jacqueline Johnson-Campbell.

Objectives

Complete administrative actions, including introductions and composition/balance discussions for committee members, and committee administrative support methodology; receive DOD briefings on the ACWA program concerning Demonstration II and engineering design studies (EDSs) status; tour and inspect the four test sites located at Aberdeen Proving Ground (APG) or the APG Edgewood Area; develop specific committee assignments for future activities; discuss concepts for planned reports and strategy; determine location and date for the next committee meeting.

Site Visit 1, June 9, 2000 (in Conjunction with Meeting 1)
Aberdeen Proving Ground Edgewood Area, Maryland, and Aberdeen, Maryland

NRC Participants

Committee Chair: Robert A. Beaudet. Committee members: Richard J. Ayen, Joan B. Berkowitz, Sheldon E. Isakoff, Frederick J. Krambeck, John A. Merson, William R. Rhyne, William R. Seeker, and Leo Weitzman. NRC staff members: Patricia P. Paulette, Bruce Braun, Harrison T. Pannella, and Jacqueline Johnson-Campbell.

Objectives

Visit the following Demonstration II test units: AEA 12-kW Silver II test unit for energetics, AEA 2 kW Silver II test unit for agent, EcoLogic gas-phase chemical reduction test unit. Visit the following EDS test units: Parsons/Honeywell immobilized-cell bioreactor test unit.

Site Visit 2, June 20–21, 2000
Dugway Proving Ground and Deseret Chemical Depot, Utah

NRC Participants

Committee Chair: Robert A. Beaudet. Committee members: Richard J. Ayen, Willard C. Gekler, Sheldon E. Isakoff, Hank C. Jenkins-Smith, John A. Merson, William R. Rhyne, Stanley I. Sandler, and Leo Weitzman. NRC staff member: Harrison T. Pannella.

Objectives

Visit the following Demonstration II test units: Teledyne-Commodore solvated-electron technology (SET) test unit for energetics, SET test unit for agent, and fluid-jet cutting test unit; Foster Wheeler supercritical water oxidation (SCWO) test unit. Visit the following EDS test units: General Atomics SCWO test unit and dunnage shredder/hydropulping system. Tour the baseline incineration-based Tooele Chemical Agent Disposal Facility and the adjacent chemical stockpile storage area.

Site Visit 3, August 2–3, 2000
Parsons, Inc., Pasadena, California

NRC Participants

Committee member: William R. Rhyne. NRC staff member: Patricia P. Paulette.

Objectives

Receive briefings on the Parsons/Honeywell technology design package. Observe working sessions of scientists and engineers discussing preliminary examinations and possible minor modifications of the package.

Committee Meeting 2, August 8–9, 2000
National Research Council, Washington, D.C.

NRC Participants

Committee Chair: Robert A. Beaudet. Committee members: Richard J. Ayen, Joan B. Berkowitz, Ruth M. Doherty, David S. Kosson, Frederick J. Krambeck, John A. Merson, William R. Rhyne, Stanley I. Sandler, William R. Seeker, and Leo Weitzman. NRC staff members: Patricia P. Paulette, Bruce Braun, Harrison T. Pannella, Jacqueline Johnson-Campbell, and Gwen Roby.

Objectives

Receive updated briefings from the ACWA technical staff, the Citizens Advisory Technical Team (CATT), and technology providers (Foster Wheeler; EcoLogic Solutions, Inc.; AEA Technology Engineering Services, Inc.; Teledyne-Commodore; Parsons). Discuss concept drafts for NRC reports on the ACWA EDS and Demonstration II; develop writing assignments; complete composition and balance; finalize locations and dates for future committee meetings and potential site visits.

Site Visit 4, August 17–18, 2000
Arthur D. Little, Inc., Cambridge, Massachusetts

NRC Participants

Committee member: Leo Weitzman. NRC staff member: Patricia P. Paulette.

Objectives

Receive briefings on the General Atomics interim engineering design technology. Observe scientific and engineering working sessions on preliminary examinations and design optimization of the package.

Site Visit 5, August 25, 2000
Illinois Institute of Technology Research Institute, Chicago, Illinois

NRC Participants

Committee member: Leo Weitzman.

Objectives

Observe Parsons CATOX unit under consideration for incorporation into the General Atomics technology package and scheduled for EDS testing.

Site Visit 6, September 6, 2000
Aberdeen Proving Ground Edgewood Area, Maryland

NRC Participants

Committee member: Leo Weitzman. NRC staff members: Patricia P. Paulette and Harrison T. Pannella.

Objectives

Attend the integrated product review meeting on the status of the engineering-scale test for the one-tenth-scale General Atomics SCWO test unit planned for treatment of VX hydrolysate at the Newport bulk storage site.

Site Visit 7, September 7, 2000
Aberdeen Proving Ground Edgewood Area, Maryland

NRC Participants

NRC staff members: Patricia P. Paulette, Bruce Braun, and Harrison T. Pannella.

Objectives

Attend the joint Program Manager for Chemical Demilitarization/Program Manager for Assembled Chemical Weapons Assessment (PMCD/PMACWA) Industry Day briefings to commence non-technology-specific acquisition RFPs activities for the disposal facility at the Pueblo Chemical Depot.

Site Visit 8, October 10, 2000
Pueblo Chemical Depot, Pueblo, Colorado

NRC Participants

NRC staff members: Patricia P. Paulette and Bruce Braun.

Objectives

Receive a briefing from the site commander of the Pueblo Chemical Depot, tour the storage bunkers, and visit the public outreach office. Attend a meeting of the Pueblo, Colorado, Citizens' Advisory Commission.

Committee Meeting 3, October 19–20, 2000
J. Erik Jonsson Center of the National Academies, Woods Hole, Massachusetts

NRC Participants

Committee Chair: Robert A. Beaudet. Committee members: Richard J. Ayen, Joan B. Berkowitz, Ruth M. Doherty, Willard C. Gekler, Sheldon E. Isakoff, David S. Kosson, Frederick J. Krambeck, John A. Merson, William R. Rhyne, Stanley I. Sandler, William R. Seeker, and Leo Weitzman. NRC staff members: Patricia P. Paulette, Bruce Braun, Harrison T. Pannella, and Chris Jones.

Objectives

Complete administrative actions, including introductions and committee composition and balance discussions. Receive updates from the PMACWA technical team and associates concerning Demonstration II, EDS, and energetics testing. Discuss initial findings and recommendations for the EDS Pueblo and Demonstration II reports.

Site Visit 9, October 25, 2000
Pine Bluff Chemical Depot, Pine Bluff, Arkansas

NRC Participants

Committee member: Willard C. Gekler.

Objectives

Observe unit operations of the SCWO reactor being tested for use in the destruction of smoke and dye materials.

Site Visit 10, October 30, 2000
Holston Army Ammunition Plant, Kingston, Tennessee

NRC Participants

Committee members: Ruth M. Doherty and William R. Rhyne.

Objectives

Attend a review of the energetics hydrolysis testing planned for Holston Army Ammunition Plant. Tour facility and observe operations. Holston is a major site for hydrolysis testing of many different types of energetic materials.

Site Visit 11, November 3, 2000
ACWA Dialogue Meeting, Pueblo, Colorado

NRC Participants

Committee Chair: Robert A. Beaudet. Committee members: Ruth M. Doherty, John A. Merson, and Hank C. Jenkins-Smith. NRC staff members: Patricia P. Paulette and Bruce Braun.

Objectives

Provide briefings to the Dialogue on NRC activities concerning the ACWA program and attend the Dialogue meeting.

Site Visit 12, November 8–10, 2000
Parsons, Inc., Pasadena, California

NRC Participants

Committee Chair: Robert A. Beaudet. Committee members: Richard J. Ayen, Willard C. Gekler, David S. Kosson (Parsons/Honeywell only), John A. Merson, William R. Rhyne, Stanley I. Sandler (General Atomics only), and Leo Weitzman. NRC staff members: Patricia P. Paulette and Harrison T. Pannella.

Objectives

Receive briefings on the Parsons/Honeywell and General Atomics EDS. Results of EDS testing were also reported.

Site Visit 13, November 26–28, 2000
Parsons, Inc., Pasadena, California

NRC Participants

Committee Chair: Robert A. Beaudet. Committee member: Willard C. Gekler.

Objectives

Attend briefings prepared for the Army ACWA team by the technology providers on the status of preliminary technical reports on the Parsons/Honeywell and General Atomics engineering designs. Results of testing for the engineering designs were also reported.

Site Visit 14, December 7, 2000
Edgewater, Maryland

NRC Participants

Committee members: Joan B. Berkowitz and Leo Weitzman.

APPENDIX C

Objectives

Attend presentation by General Atomics to PMACWA concerning the developmental status of its SCWO reactor.

Committee Meeting 4, December 14–15, 2000
National Academy of Sciences, Washington, D.C.

NRC Participants

Committee Chair: Robert A. Beaudet. Committee members: Richard J. Ayen, Joan B. Berkowitz, Ruth M. Doherty, Willard C. Gekler, Sheldon E. Isakoff, Hank C. Jenkins-Smith, David S. Kosson, Frederick J. Krambeck, John A. Merson, William R. Rhyne, Stanley I. Sandler, William R. Seeker, and Leo Weitzman. NRC staff members: Patricia P. Paulette, Harrison T. Pannella, and Jacqueline Johnson-Campbell.

Objectives

Report development; review and revise current versions of EDS Pueblo and Demonstration II reports; devise path forward for first full message drafts of each report.

Site Visit 15, January 4, 2001
Holston Army Ammunition Plant, Kingsport, Tennessee

NRC Participants

Committee members: Ruth M. Doherty, John A. Merson, and William R. Rhyne.

Objectives

Attend a working review meeting on energetics hydrolysis testing at Holston and the results of this testing to date. Note the scientific and engineering issues revealed during the test procedures.

Site Visit 16, January 24, 2001
ACWA Dialogue Meeting, Lexington, Kentucky

NRC Participants

Committee member: William R. Rhyne. NRC staff member: Patricia P. Paulette.

Objectives

Attend Dialogue meeting and provide an update of the NRC ACW II Committee's activities.

Committee Meeting 5, February 8–9, 2001
Washington, D.C.

NRC Participants

Committee Chair: Robert A. Beaudet. Committee members: Richard J. Ayen, Joan B. Berkowitz, Ruth M. Doherty, Willard C. Gekler, David S. Kosson, Frederick J. Krambeck, John A. Merson, William R. Rhyne, Stanley I. Sandler, William R. Seeker, and Leo Weitzman. NRC staff members: Patricia P. Paulette, Bruce Braun, Harrison T. Pannella, and Jacqueline Johnson-Campbell.

Objectives

Receive briefings from the PMACWA technical staff concerning the status of EDS I, EDS II, and Demonstration II program activities and other presentations from the energetics hydrolysis testing group at Picatinny Arsenal. Review and revise reports. Define steps to EDS I report concurrence draft. Set goals for interim activities and for the next meeting.

Information-Gathering Activity, March 11–14, 2001
Houston, Texas

NRC Participants

NRC staff member: Patricia P. Paulette.

Objectives

Attend National Association of Corrosion Engineers symposium on SCWO technology as it relates to materials of construction and corrosion reactions in reactors. Arrange to obtain scientific briefing packages and technical papers for distribution to committee members.

Committee Meeting 6, March 26–27, 2001
Beckman Center, Irvine, California

NRC Participants

Committee Chair: Robert Beaudet. Committee members: Richard J. Ayen, Joan B. Berkowitz, Ruth M. Doherty, Willard C. Gekler, Sheldon E. Isakoff, Frederick J. Krambeck, John A. Merson, William R. Rhyne, Stanley I. Sandler, William R. Seeker, and Leo Weitzman. NRC staff members: Bruce Braun, Patricia Paulette, Harrison Pannella, Gwen Roby, and William Campbell.

Objectives

Closed meeting for report development. Complete concurrence draft for EDS I Pueblo report and preconcurrence draft for Demonstration II report. Schedule next meeting date and discuss items for the agenda.

Appendix D

Biographical Sketches of Committee Members

Robert A. Beaudet, *Chair*, received his Ph.D. in physical chemistry from Harvard University in 1962. From 1961 to 1962, he was a U.S. Army officer and served at the Jet Propulsion Laboratory as a research scientist. He joined the faculty of the University of Southern California in 1962 as an assistant professor and was chair of the Chemistry Department from 1976 to 1979. He has also served on Department of Defense committees addressing chemical warfare agents in both offensive and defensive scenarios. He was chair of an Army Science Board committee that addressed chemical detection and trace-gas analysis and chair of an Air Force technical conference on chemical warfare decontamination and protection. He has served on two National Research Council (NRC) studies on chemical and biological sensor technologies and energetic materials and technologies. Most of his career has been devoted to research in molecular structure and molecular spectroscopy. Dr. Beaudet previously served as a member of the Board of Army Science and Technology (BAST) and as a BAST liaison to the Committee on Review and Evaluation of the Army Chemical Stockpile Disposal Program (Stockpile Committee), a standing NRC committee. He is currently a member of the NRC Committee on Review of the Non-Stockpile Chemical Materiel Disposal Program. Dr. Beaudet is the author or coauthor of more than 100 technical reports and papers.

Richard J. Ayen received his Ph.D. in chemical engineering from the University of Illinois. Dr. Ayen is a former vice president of technology for Waste Management, Inc., and is now an independent consultant. He has extensive experience in the evaluation and development of new technologies for the treatment of hazardous waste. Dr. Ayen managed all aspects of the Waste Management Clemson Technical Center, including treatability studies and technology demonstrations for hazardous and radioactive waste. He has published extensively in his fields of interest and is a member of the NRC Committee on Review of the Non-Stockpile Chemical Materiel Disposal Program.

Joan B. Berkowitz, who graduated from the University of Illinois with a Ph.D. in physical chemistry, is currently managing director of Farkas Berkowitz and Company. Her areas of expertise include environmental and hazardous waste management, available technologies for the cleanup of contaminated soils and groundwater, and physical and electrochemistry. She has contributed to several studies by the Environmental Protection Agency, been a consultant on remediation techniques, and assessed various destruction technologies. Dr. Berkowitz has written numerous publications on hazardous waste treatment and environmental subjects. She is currently a member of the NRC Committee on Review of the Non-Stockpile Chemical Materiel Disposal Program.

Ruth M. Doherty received a Ph.D. in physical chemistry from the University of Maryland. She is currently technical advisor for the Energetic Materials Research and Technology Department, Naval Surface Warfare Center, Indian Head, Maryland. Since 1983, she has coauthored about 60 publications on physical chemistry. In the past 6 years, Dr. Doherty has given 20 presentations on various aspects of the science and technology of explosives. In 1998 and 1999, she delivered a series of lectures on explosives technology for members of the Office of Naval Intelligence. For more than 15 years, she has been involved in research and development of energetics materials and explosives at the Naval Surface Warfare Center.

Willard C. Gekler graduated from the Colorado School of Mines with a B.S. in petroleum refining engineering and pursued additional graduate study at the University of California in Los Angeles in nuclear engineering. Mr. Gekler is currently an independent consultant working for his previous employer, EQE International, Inc. His extensive experience includes membership on the NRC ACW I and II Committees and on the Expert Panel reviewing the quantitative risk assessments and safety analyses of hazardous materials handling, storage, and waste treatment systems for the Anniston and Umatilla chemical disposal facilities. His expertise is in hazard evaluation, quantitative risk analyses, reliability assessment, and database development for risk and reliability. Mr. Gekler is a certified reliability engineer and a member of the Society for Risk Analysis (SRA) and the American Nuclear Society. He is author or coauthor of numerous publications.

Sheldon E. Isakoff, who received his Ph.D. in chemical engineering from Columbia University, is the retired director of the Engineering R&D Division of E.I. du Pont de Nemours and Company. His experience includes the management of technology, directing research and development, market assessment and development, process scale-up, commercial introduction, and leadership of personnel. His areas of expertise also include materials science and engineering and the development and application of new materials for industrial and consumer markets. Dr. Isakoff is a fellow and past president of the American Institute of Chemical Engineering and a former director of its materials engineering and sciences division. He was elected to the National Academy of Engineering in 1980 and has served on several NRC committees.

Hank C. Jenkins-Smith received his Ph.D. in political science from the University of Rochester and is currently a professor in the Department of Political Science at the University of New Mexico (UNM). He is also the director of the UNM Institute for Public Policy. His areas of expertise include statistical analysis, measurement of public opinion, politics of risk perception, environmental policy, and public policy. Dr. Jenkins-Smith is a member of the Society for Risk Analysis (SRA) and the American Political Science Association. In 1996, he received the SRA's Risk Research Award. He is the author of more than 60 publications and reports.

David S. Kosson has a B.S. in chemical engineering, an M.S. in chemical and biochemical engineering, and a Ph.D. in chemical and biochemical engineering from Rutgers, the State University of New Jersey. He is chairman and professor of the Department of Civil and Environmental Engineering and professor of chemical engineering at Vanderbilt University and a former professor of chemical and biochemical engineering at Rutgers. Dr. Kosson has carried out research and published extensively on subsurface contaminant transport phenomena; leaching phenomena; physical, chemical, and microbial treatment processes for hazardous waste; and waste management policy. Dr. Kosson served on the NRC Committee on Review and Evaluation of the Army Stockpile Disposal Program for 7 years, the final 2 years as chair. As a member of the NRC Committee on Alternative Chemical Demilitarization Technologies and the Panel on Review and Evaluation of Alternative Chemical Disposal Technologies, he contributed to the Army's decision to use alternative methods of destruction at both the Aberdeen and Newport facilities. Dr. Kosson is well known for his expertise in bioremediation.

Frederick J. Krambeck received his Ph.D. in chemical engineering from the City University of New York. He is a senior consultant for ExxonMobil Research and Engineering Company. His expertise includes research and development (R&D) in petroleum refining, including process and reactor design and development, chemical reaction engineering, on-line and off-line optimization, modeling, and R&D project management. He is also experienced in technology strategy considerations for greenhouse gas stabilization. Dr. Krambeck was elected to the National Academy of Engineering in 1999 and is a fellow and member of the Board of Directors of the American Institute of Chemical Engineers (AIChE). He has assisted in the development of patents for more than 25 processes. Dr. Krambeck is the author or coauthor of 40 publications, including *Elements of Process Engineering,* which was delivered as a plenary lecture for the AIChE 90th Anniversary History Session in 1998.

John A. Merson received a B.S. and M.S. in chemical engineering from the University of New Mexico and a Ph.D. in chemical engineering from Arizona State University. His areas of expertise include research, development, and application of energetic materials and components in the nuclear weapons stockpile. Dr. Merson is the department manager of the Explosive Subsystems and Materials Department at Sandia National Laboratories, which designs, develops, and characterizes explosive, propellant, and pyrotechnic components and subsystems to meet specific needs. Dr. Merson is a member of the American Institute of Chemical Engineers.

William R. Rhyne received a B.S. in nuclear engineering from the University of Tennessee and an M.S. and D.Sc. in nuclear engineering from the University of Virginia and is cofounder and director of H&R Technical Associates, Inc. Dr. Rhyne has extensive experience in risk and safety analysis associated with nuclear and chemical processes and with the transport of hazardous nuclear materials and chemicals. From 1984 to 1987, he was the project manager and principal investigator for a probabilistic accident analysis of transporting obsolete chemical munitions. Dr. Rhyne is the author or coauthor of more than 40 publications and reports on nuclear and chemical safety and risk analysis, including *Hazardous Materials Transportation Risk Analysis: Quantitative Approaches for Truck and Train.* He is a member of the NRC Transportation Research Board Hazardous Materials Committee, the Society for Risk Assessment, the American Nuclear Society, and the American Institute for Chemical Engineers.

Stanley I. Sandler, who received his Ph.D. in chemical engineering from the University of Minnesota, is currently the

Henry Belin du Pont Chair and director of the Center for Molecular and Engineering Thermodynamics at the University of Delaware. His extensive research interests include applied thermodynamics and phase equilibrium, environmental engineering, and separations and purification. Dr. Sandler is a recipient of the Warren K. Lewis Award from the American Institute of Chemical Engineers and the Inaugural E.A. Mason Memorial Lecturer Award from Brown University. He is a member of the National Academy of Engineering and has published more than 250 technical articles in recognized journals and conference proceedings.

William R. Seeker received his Ph.D. in engineering (nuclear and chemical) from Kansas State University. He is senior vice president and member of the Board of Directors of the Energy and Environmental Research Corporation, a wholly owned subsidiary of General Electric Company. He has extensive experience in the use of thermal treatment technologies, environmental control systems for managing hazardous waste, and air pollution control. He is a member of the Executive Committee of the EPA Science Advisory Board and the author of more than 150 technical papers on various aspects of technology and the environment.

Leo Weitzman received his Ph.D. in chemical engineering from Purdue University. He is a consultant with 28 years of experience in the development, design, permitting, and operation of equipment and facilities for the treatment of hazardous wastes and remediation debris. Dr. Weitzman has extensive experience in the disposal of hazardous waste and contaminated materials by thermal treatment, chemical reaction, solvent extraction, biological treatment, and stabilization. He has published more than 40 technical papers.